Regional Landscape Conservation Planning

Edited by Reed F. Noss
and Thomas S. Hoctor

REGIONAL LANDSCAPE CONSERVATION PLANNING

Wildlife, Connectivity, and the Florida Model

University of Florida Press Gainesville

Cover image: A female Florida panther and her kittens trigger a camera trap in Audubon's Corkscrew Swamp Sanctuary near Naples. The recovery of the Florida panther depends on the protection of the Florida Wildlife Corridor to allow the breeding population of panthers to reclaim historic territory throughout the state. Photo courtesy of Carlton Ward Jr. / Wildpath. *Cover design*: Katrina Noble

Publication of this work was made possible by funding from the Florida State Government supporting the science and protection of the Florida Ecological Greenways Network and the Florida Wildlife Corridor.

Published in the United States of America

31 30 29 28 27 26   6 5 4 3 2 1

DOI: http://doi.org/10.5744/9781683406549

Library of Congress Cataloging-in-Publication Data
Names: Noss, Reed F. editor | Hoctor, Tom (Thomas) editor
Title: Regional landscape conservation planning : wildlife, connectivity, and the Florida model / edited by Reed F. Noss and Thomas S. Hoctor.
Description: Gainesville : University of Florida Press, 2026. | Includes bibliographical references and index.
Identifiers: LCCN 2026008381 (print) | LCCN 2026008382 (ebook) | ISBN 9781683406549 hardback | ISBN 9781683405962 ebook | ISBN 9781683405832 pdf
Subjects: LCSH: Landscape protection—Florida—Planning | Wildlife conservation—Florida—Planning | Nature conservation—Florida—Planning | Landscape protection—Planning | Nature conservation—Planning | BISAC: SCIENCE / Life Sciences / Ecology | ARCHITECTURE / Landscape
Classification: LCC QH76.5.F6 R45 2026 (print) | LCC QH76.5.F6 (ebook)

LC record available at https://lccn.loc.gov/2026008381
LC ebook record available at https://lccn.loc.gov/2026008382

UF PRESS

University of Florida Press
PO Box 140239
Gainesville, FL 32614
floridapress.org

GPSR EU Authorized Representative: Mare Nostrum Group B.V., Doelen 72, 4831 GR Breda, The Netherlands, gpsr@mare-nostrum.co.uk

Contents

List of Figures vii
List of Tables ix

Introduction
Reed F. Noss and Thomas S. Hoctor
1

1 Conservation Planning in a Biodiversity Hotspot
Reed F. Noss and Thomas S. Hoctor
5

2 The Influences, Precursors, and Early History of Regional Landscape Conservation Planning
Reed F. Noss and Thomas S. Hoctor
22

3 Prioritization Methods for Ambitious Land Conservation under Preservation 2000, Florida Forever, and the Rural and Family Lands Protection Program
Jon Oetting, Amy Knight, and Thomas S. Hoctor
46

4 Statewide Land-Use Planning and Growth Modeling
Margaret (Peggy) Carr, Richard Doty, Daniel Farrah, Jason Teisinger, Liz Thompson, Michael Volk, and Paul Zwick
68

5 Identifying and Protecting Climate Refugia at Multiple Spatial Scales
Eve Bohnett, Reed F. Noss, Alan Flint, Lorraine Flint, Michael O'Brien, Paul Donsky, Isabella Guttuso Browne, Derya Yesilkusak, and Thomas S. Hoctor
86

6 Applying a Landscape Approach to Aquatic and Coastal Conservation
Aaron Adams, JoEllen Wilson, Philip Stevens, and Emily Taylor
109

7 The Florida Ecological Greenways Network and Florida Wildlife Corridor: Protecting Biodiversity and Ecosystem Services
Thomas S. Hoctor, Michael O'Brien, Carlton Ward Jr., Mallory Dimmitt, Joseph M. Guthrie, Eve Bohnett, Sarah Lockhart, Asena Goren, Reed F. Noss, Jason Lauritsen, Marly Fuller, Kailyn Lawson, and Derya Yesilkusak
130

8 Road Ecology Planning and Practice
Daniel J. Smith
156

9 How Land Trusts Implement Landscape Conservation
Susan Carr, Reed F. Noss, Sarah Lockhart, Ramesh Buch, Shane Wellendorf, Debi Osborne, Ken Grudens, Lauren Day, Kimberly Tillman, and Tom Kay
182

10 Florida's Paradoxical Influence on Wildlands Protection and Restoration
John Davis, Jodi Hilty, Harvey Locke, Ron Sutherland, Charles Chester, Sonam Wangdi, and Katie Shepard
209

Acknowledgments 233
List of Contributors 235
Index 247

Figures

1.1 Summed range-size rarity in the southeastern United States for Critically Imperiled, Imperiled, and ESA-listed species of plants, vertebrates, and several invertebrate groups 9
2.1 Models for human-dominated landscapes and relatively undeveloped landscapes 35
2.2 The first proposed Florida Wildlife Corridor network 37
3.1 Result of 1990s mapping charette 49
3.2 Areas of Conservation Interest / Potential Natural Areas 50
3.3 Original Strategic Habitat Conservation Areas 51
3.4 Detail of Significant Surface Waters priorities map 56
3.5 Detail of Greenways Florida Forever Strategic Priorities map in North Florida 59
4.1 Decennial population growth in Florida, 1920–2020 69
5.1 The Atlantic Warm Pool 91
5.2 National climate projections for global warming 92
5.3 Climate refugia for maximum temperature in Florida 94
5.4 Florida springs and selected springsheds 97
5.5 Sea-level rise projections for Florida 99
7.1 The process to delineate the FEGN boundary 132
7.2 The 2021 FEGN conservation priorities 137
7.3 Additions to the FEGN network since 2016 137
8.1 A wildlife technician documenting a white-tailed deer involved in a vehicle collision 158
8.2 Roadside hair snare placement and a successful snag 159
8.3 Locations of frequent road crossings by black bears 160
8.4 The relationship between landform, road curvature, and location of animal-vehicle collisions 161

8.5 An array of wildlife crossing structure designs employed in dry and wet environments 163
8.6 Location of panther-vehicle collisions on Florida roads (1998 to 2022) 168
8.7 Location of bear-vehicle collisions on Florida roads (1998 to 2022) 168
8.8 Roadless areas in Florida 170
8.9 Existing and proposed wildlife crossing structures in Florida 170
8.10 Large animal-vehicle collisions identified in the Florida Wildlife Corridor 173
8.11 Florida panther movements and potential road crossing locations within south-central Florida 174
9.1 Current list of land trusts operating in Florida 185
9.2 Case study areas in the context of the Florida Wildlife Corridor 202
10.1 System of protected and conserved areas and corridors of Bhutan 217

Tables

4.1 Comparison of major modeling assumptions for 2070 Trend and 2070 Alternative 77

4.2 Acreage comparison of Florida 2070 alternative population allocation scenarios 77

4.3 Population projections used for Sea Level 2040/2070 80

6.1 Summary of Florida agencies responsible for governance of water resources 116

7.1 Priority Ecological Area criteria from the 2021 FEGN 133

7.2 Land category statistics for the 2021 FEGN boundary 138

Regional Landscape Conservation Planning

Introduction

Reed F. Noss and Thomas S. Hoctor

Many books have been written on nature conservation and quite a few on conservation planning. In the academic journals devoted to conservation biology and the protection and management of natural areas, conservation planning has been one of the most frequently covered topics for more than three decades. Hundreds of articles are published each year on conservation planning, making it one of the most popular subdisciplines of conservation biology. Out of this vast literature several interesting patterns have emerged. One striking pattern is that, especially in the early years, science-based conservation planning was dominated by Australians, followed increasingly over the years by authors from other developed countries. These are subjective impressions; we haven't analyzed the statistics. Three authors from Australia recently conducted such a study, however. Cobb et al. (2024) performed a bibliometric analysis of the geospatial conservation planning literature and found that "westernized nations dominated the field, with the United States, the United Kingdom, and Australia being responsible for almost two-thirds of the research globally, with research interest exponentially growing since 2010." Of the top 25 countries that have contributed publications on conservation planning, only six are outside of North America and Europe (Cobb et al. 2024).

Some environmentalists have criticized conservation planning, and conservation biology in general, for being dominated by scientists and planners from westernized nations, whereas most biodiversity is in developing countries, especially in the tropics. That's only a partly legitimate criticism. It is true that the highest known richness of plant and animal species is in tropical forests. However, of the 36 currently recognized global biodiversity hotspots, each of which holds at least 1,500 endemic plant species and has suffered at least 70% loss of primary habitat, 19 are tropical and 17 are entirely or mostly outside the tropics (Critical Ecosystem Partnership, undated). One of those mostly outside the tropics is the North American Coastal Plain, within which Florida is the most biodiverse portion (see chapter 1).

Of course, protection of nature is important for many reasons in every region of the world, not just in hotspots. For instance, conservation is now recognized to include ecosystem service provision and food security. Among the most critical ecosystem services are the direct benefits nature provides to people by enhancing physical and psychological health and intellectual development (Flies et al. 2017; Aerts et al. 2018). Also increasingly recognized are the interrelationships between biodiversity conservation and rural land uses, including forestry and agriculture, and the well-being of the often-marginalized people employed in these industries. By protecting and managing prudently large, interconnected landscapes, we can provide for nature and for humans.

Florida has been a leader in science-based and map-based conservation planning for several decades. Indeed, Florida is a precedent-setter and inspiration for large-scale conservation planning worldwide. The planning concepts applied and the lessons learned in landscape conservation in Florida have influenced and continue to influence conservation planning globally. What we call in this book the "Florida model" is one of the first applications—apparently the very first—of a now common regional planning design consisting of core reserves, buffer (or multiuse) zones, and ecological corridors. It basically combines the UNESCO biosphere reserve model, and its more universal derivative, the multiple-use module (Harris 1984), with the empirically validated tool of wildlife movement corridors.

We are not producing this book merely to showcase the Florida model, though admittedly that is part of our motivation because we are unabashedly proud of it. More important, in the big scheme of things, is providing a tangible example of how an extraordinarily ambitious and initially controversial design became popular with the public and politicians and was implemented through generous funding for conservation land acquisition. The design is not yet fully implemented, but we are well on our way. It is our hope that the empirical lessons and methodologies described in this book can be applied by conservation planners elsewhere to their own regions, including expanding current conservation partnerships beyond Florida's borders to adjoining states in the southeastern US and beyond through the proposed Eastern Wildway (see chapter 10).

A fundamental lesson of the Florida experience is that no idea or plan for conservation should ever be rejected as too radical, impractical, or unrealistic. The history of regional landscape conservation planning in Florida and now elsewhere clearly demonstrates that ideas once scoffed at can ultimately become well-accepted. And then, somewhat magically, they are implemented (at least in part). On a related theme, most people outside Florida are astounded that a conservative southern state would spend so much money on conservation land acquisition. But conservation in Florida has almost always been a bipartisan endeavor, widely supported by the public. Most people are also surprised that Florida is a biodiversity hotspot and that it still has significant wildland and biodiversity left to protect. We hope to keep surprising those folks.

We selected the chapters in this book to illustrate a broad spectrum of conservation planning topics, methods, and challenges. Collectively, the chapters cover Florida's extreme biological richness (chapter 1), the precursors and history of regional landscape conservation planning in Florida and elsewhere (chapters 2, 3, 7, and 10), prioritization methods (chapter 3), and statewide land-use planning and growth modeling (chapter 4). We explore the importance of climate refugia and how they can be identified and protected (chapter 5). We include a contribution on the application of a landscape approach to aquatic and coastal conservation (chapter 6), a topic usually ignored in conservation planning.

We review the history and contributions of the widely acclaimed Florida Ecological Greenways Network and Florida Wildlife Corridor (chapter 7), the critical topic of road ecology as it relates to wide-ranging species and large landscape conservation (chapter 8), and the key role of land trusts and private lands in conservation (chapter 9). The book closes with a look at some other major regional conservation initiatives in North America and beyond that have been directly or indirectly inspired by the Florida model (chapter 10). We elected not to delve too deeply into conservation policy, politics, or socioeconomics in Florida or elsewhere, as those issues have been explored by others (e.g., Henderson 2022). We also do not wade into the technical details of site-selection algorithms for conservation prioritization (but see chapter 3 for use of the Marxan algorithm) or of ecological monitoring and adaptive management. Most of all, this book endeavors to show what can be accomplished in conservation when an idealistic, unfaltering vision is combined with sound science and practice informed by experience.

Literature Cited

Aerts, R., O. Honnay, and A. Van Nieuwenhuyse. 2018. Biodiversity and human health: Mechanisms and evidence of the positive health effects of diversity in nature and green spaces. British Medical Bulletin 127:5–22.

Cobb, G., J. Nalau, and A. L. M. Chauvenet. 2024. Global trends in geospatial conservation planning: A review of priorities and missing dimensions. Frontiers in Ecology and Evolution. https://doi.org/10.3389/fevo.2023.1209620.

Critical Ecosystem Partnership Fund. Undated. Hotspots: North American Coastal Plain. https://www.cepf.net/our-work/biodiversity-hotspots/north-american-coastal-plain. Accessed July 30, 2024.

Flies, E. J., C. Skelly, S. Singh Negi, P. Proabhakaran, Q. Liu, K. Liu, F. C. Goldizen, C. Lease, and P. Weinstein. 2017. Biodiverse green spaces: A prescription for global urban health. Frontiers in Ecology and the Environment 15:510–516.

Harris, L. D. 1984. The Fragmented Forest. Island Biogeographic Theory and the Preservation of Biotic Diversity. University of Chicago Press, Chicago.

Henderson, C. 2022. Forces of Nature: A History of Florida Land Conservation. University Press of Florida, Gainesville.

1

Conservation Planning in a Biodiversity Hotspot

Reed F. Noss and Thomas S. Hoctor

Introduction

Conservation planning is a subdiscipline of conservation biology, a science in the service of conservation (Noss et al. 1997). As a crisis discipline (Soulé 1985; Meine et al. 2006), conservation biology is meant to be applied to the real world rapidly. However, it has not been applied quickly enough or often enough, or we would not be facing a global biodiversity cataclysm. We would not be well into the sixth mass extinction event in the history of life (Ceballos and Ehrlich 2023). Knight et al. (2008) found that "two-thirds of conservation assessments published in the peer-reviewed scientific literature do not deliver conservation action, primarily because most researchers never plan for implementation." This science-implementation gap is a serious problem. It almost appears that conservation planners have been wasting their time on a pointless academic exercise.

Given the alarming rate of urban, suburban, and exurban development in Florida and other high-growth regions, it is crucial that conservation assessments and plans be implemented while opportunities for land protection remain. It might seem that the odds are stacked against achieving meaningful conservation in Florida, given the progrowth political climate and the relentless domestic and foreign immigration pressure—more than 1,000 new residents daily (Sarasota Herald-Tribune 2024). Yet paradoxically, Florida offers an optimistic vision for the world of a place where the science-implementation gap has been significantly bridged and where ambitious goals for land conservation are well on their way to achievement. The broader lesson is that if we can accomplish regional landscape conservation in Florida, it can be done almost anywhere there is a will of the people and rural land left to save or restore.

Conservation action is especially imperative in biodiversity hotspots. Biodiversity is not distributed randomly or evenly across the Earth's surface. It is concentrated in particular places known as hotspots. The urgency and challenges of conservation planning are magnified in biodiversity hotspots because the costs of failure are so dire. Basically, a hotspot is a region with a lot of threatened biodiversity. Globally, hotspots are recognized as distinct biogeographic regions that, on the positive side, contain at least 1,500 endemic plant species (i.e., found nowhere else) but, on the negative side, have suffered at least 70% loss of their primary vegetation (Myers et al. 2000; Mittermeier et al. 2011). Global hotspots also usually hold high concentrations of endemic animals, though plants provide the standard for comparison. Protecting these threatened centers of endemism may be the most cost-efficient way to reduce global extinction rates.

The hotspots approach is an example of the more general and eminently logical irreplaceability-vulnerability framework for prioritizing sites or regions for conservation (Margules and Pressey 2000). Irreplaceability is a measure of a site's or region's contribution to conservation goals or, alternately, the extent to which those goals will be compromised if that area is not protected (Pressey et al. 1994). Vulnerability is the relative risk of biodiversity loss and is normally higher in regions with a history of massive habitat destruction, which are closer to a threshold of ecosystem collapse and population extinctions, as well as in regions with rapidly escalating threats such as human population growth. The irreplaceability-vulnerability framework has implicitly or explicitly guided global and regional conservation planning for decades, including the World Wildlife Fund's Global 200 ecoregions project (Olson and Dinerstein 1998; Olson et al. 2001) and many regional conservation plans (e.g., Noss et al. 2002; Cowling et al. 2003; Kremen et al. 2008) as well as the biodiversity hotspots strategy by that name.

In this chapter we review Florida's status as a hotspot of irreplaceability and vulnerability and discuss what that means for conservation planning—a discussion that is equally relevant to other highly biodiverse and rapidly developing regions. The chapters in this book mostly focus on the Florida model of conservation planning, one in which the entire regional landscape—not only the wildest remaining lands but also the seminatural matrix of working ranches and timberlands and even the green places within cities—is examined for its potential contributions to conservation. Big natural and seminatural areas connected by broad corridors designed for panthers, bears, and other wide-ranging animals (e.g., the Florida Wildlife Corridor; see chapter 7) get the most attention because opportunities to protect these wildlands and their dependent species are fading quickly. Big animals, especially carnivores, and wilderness enthuse the public: the Florida panther is the state's official animal, serves as the mascot for a Florida professional ice hockey team, and is unfailingly popular among Floridians. And these large landscapes are equally significant for providing ecosystems services like protection from flooding (which is a huge threat in Florida).

Nevertheless, many of Florida's endemic species, particularly plants and invertebrates, as well as climate microrefugia such as caves, sinkholes, and springs (see chapter 5) are small and localized and require a fine-scale approach. Appropriate conservation strategies for these species and features are often quite different from those focused on wide-ranging animals or large-scale ecosystem processes. A comprehensive regional landscape approach to conservation (Noss 1983) addresses conservation needs at all relevant spatial and temporal scales and recognizes the complementarity of large and small reserves. It is not one or the other.

The Florida Hotspot: Irreplaceable and Vulnerable

IRREPLACEABILITY

Florida lies within the North American Coastal Plain, which is geologically and floristically sharply delineated (Fenneman 1938; Takhtajan 1986). The North American Coastal Plain was recognized as the thirty-sixth global biodiversity hotspot in 2016 after biologists documented that it met official criteria for endemism and extent of habitat loss (Noss et al. 2015; Critical Ecosystem Partnership Fund undated). At approximately the same time, the southeastern portion of the Coastal Plain was recognized as one of the major global hotspots for turtles (Mittermeier et al. 2015).

Notably, Florida is the hottest spot biologically within the Coastal Plain hotspot and the entire southeastern US, with many concentrated occurrences of rare, restricted-range, and federally listed species (fig. 1.1); it is also one of the richest regions for rare and endemic species in North America (Hamilton et al.

2022). Four decades ago, Gentry (1986) showed that Florida has 385 endemic plant species, a number second only to California (with 1,517), and pointed out how surprising this high endemism is for an essentially flat and relatively homogeneous region. The current estimate of endemic or near-endemic (>90% of distribution) plant taxa in Florida is 554, with a strong concentration in xeric sand habitats (Alan Weakley, University of North Carolina Herbarium, pers. comm.). Particularly interesting, a recent phylogenetic analysis shows that Florida stands out in North America—and is well above California—in having a rich mix of rare long-branch (paleoendemic) and rare short-branch (neoendemic) plant species (Mishler et al. 2020). This combination suggests that Florida has been both an evolutionary cradle (where species are "born") and an evolutionary museum (where species are retained) (Stebbins and Major 1965). Florida also has the highest number of plant families in North America north of Mexico, another indication of deep diversity (BONAP, undated).

Florida's high richness of both paleoendemic and neoendemic species is closely tied to its status as a long-term climate-change refugium (see chapter 5). Probably all of Earth's centers of endemism have served as long-term climate refugia (Harrison and Noss 2017). The Coastal Plain is no exception, and Florida, as mostly a peninsula buffered on three sides by predictably mild ocean currents, has shown the greatest refugial capacity.

The climate refugium role of Florida most obviously explains the prevalence of paleoendemics—they had a relatively stable environment in which to survive for up to tens of millions of years. But a relatively stable climate also promotes the survival of recently evolved neoendemics (Harrison and Noss 2017). According to the well-accepted model of allopatric speciation (Mayr 1942; Dobzhansky 1951), virtually all species start out as small, isolated populations, and such populations are more likely to go extinct from stochastic processes (Lande 1993). In a relatively stable environment, these processes are ameliorated, and the probability of persistence is enhanced. Many of Florida's neoendemics, especially in the peninsula, apparently speciated on islands (now ridges) isolated by high sea-level stands at various times from the mid-Miocene through the Pleistocene—that is, from around 10 million to 12,000 years ago (James 1961; Sorrie and Weakley 2001; Germain-Aubrey et al. 2014). Neoendemic plants and animals are also concentrated in the Florida Keys; we know these species are young because the Keys have been isolated from one another and the mainland for only some 6,000 years, and none of these species is known to have occurred on the mainland (Noss 2012).

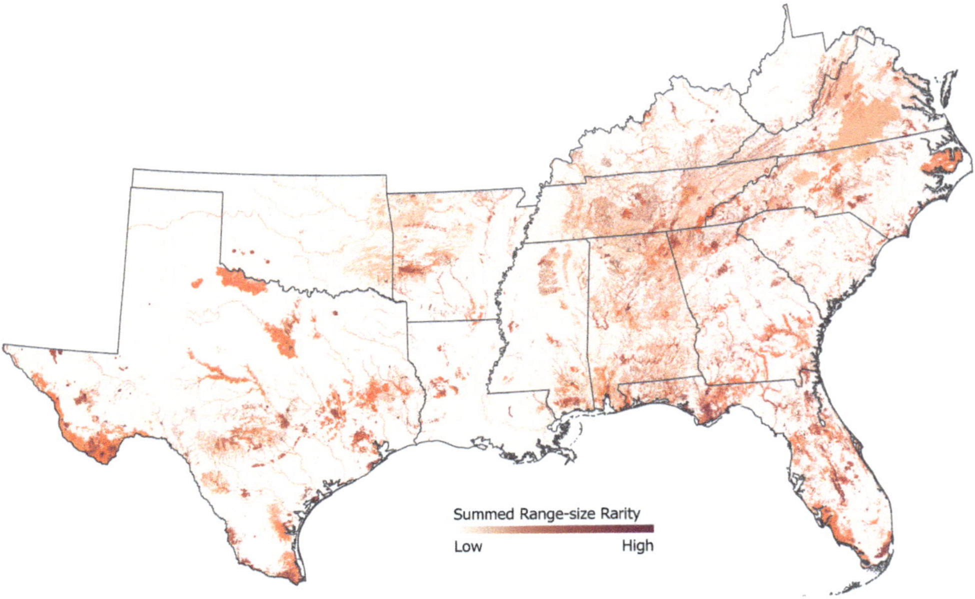

FIGURE 1.1. This map depicts summed range-size rarity in the southeastern United States for Critically Imperiled (categorized by NatureServe as "G1"), Imperiled ("G2"), and ESA-listed species (i.e., species listed as Endangered or Threatened under the Endangered Species Act) in the following groups: vertebrates, freshwater mussels, crayfishes, bumblebees, solitary bees, butterflies, skippers, and vascular plants. Darker shades of red identify where species with very small ranges (and thus fewer places where they can be conserved) are likely to occur or where the presence of multiple imperiled species contributes to higher scores. From NatureServe Network (2024). ©NatureServe, all rights reserved.

Most of Florida's native species did not evolve here but rather colonized Florida from other regions. Florida's complicated biogeography reflects contributions to its flora and fauna from many regions, including the Appalachians and Piedmont, the West Indies, the Great Plains and Midwest, and most intriguing, the American Southwest (and, for some taxa, southward to Central America) (Noss 2012). Among the many western or neotropical animal taxa that have disjunct species or populations in Florida, and in some cases other southeastern states, are tortoises (*Gopherus*), scrub-jays (*Aphelocoma*), indigo snakes (*Drymarchon*), caracaras (*Caracara*), white-tailed kites (*Elanus*), burrowing owls (*Athene*), harvester ants

(*Pogonomyrmex*), and several other ants, bees, and wasps (Hymenoptera). Western plants with disjunctions in Florida include grama grass (*Bouteloua*), wild buckwheat (*Eriogonum*), yuccas (*Yucca*), ziziphus (*Pseudoziziphus*), and various cacti (Cactaceae). In the recent (Plio-Pleistocene) past, Florida had jackrabbits, pronghorns, ground squirrels, pallid bats, condors, teratorns, prairie-chickens, and other species associated with the American West (Noss 2012). It is thought that many species colonized Florida from the west along a Gulf Coastal Corridor, a stretch of savanna and scrub habitats that existed intermittently but over long periods of time from the mid-Miocene through at least the early Pleistocene (Webb 1977; Morgan and Emslie 2010).

Another factor explaining Florida's high biodiversity is its active disturbance regime, particularly fire, wind, and floods. Some intermediate to rather high level of natural disturbance tends to promote high species richness (Connell 1978; Huston 1979). Approximately 85% of the plant species endemic to the Coastal Plain are associated with fire-dependent pine savannas and embedded ecosystems, particularly depression wetlands (Noss et al. 2015). Florida is one of the most fire-prone regions of the world; based on fossil evidence, it has been for millions of years (Noss 2018). Most of our endemic terrestrial and wetland species and natural communities are fire dependent and would be eliminated in the absence of fire. The *General Map of Natural Vegetation of Florida* (Davis 1967) shows roughly 90% of the land surface of the state dominated by fire-prone vegetation prior to Euro-American settlement.

Fire and other natural disturbances reduce competitive exclusion among species by reducing biomass and exposing bare soil for colonization by plants (Grime and Pierce 2012). A heterogeneous fire regime, especially with respect to fire intensity, maintains high plant-species richness "by enhancing individual performance (growth, survival, and fecundity) of rare recruits in sites with low densities of competitors . . . as well as by promoting coexistence of species with different postfire regeneration traits" (Myers and Harms 2011). High richness of endemic species, especially paleoendemics, in Florida's fire-dependent ecosystems suggests great antiquity of these ecosystems and refutes popular but uninformed notions (e.g., Mann 2005; Pyne 2016) that Florida is very young geologically and that fires here were entirely set by humans (Noss 2018). Regional landscape conservation planning is essential for fire management because, among other things, prescribed burning is challenging to impossible in a highly fragmented landscape where small patches of natural habitat are surrounded by urban land use—this is another primary reason for pursuing protection of large, intact landscapes across the state.

VULNERABILITY

Florida's biodiversity is rich and irreplaceable, but it is also highly vulnerable. Florida is experiencing the highest rate of habitat loss in the Southeast and likely the nation, inextricably tied to its terrifyingly high rate of human population growth, which surpasses that of all other states (US Census Bureau 2022; see chapter 4). This growth is a relatively new phenomenon in Florida, intensifying only since the mid-twentieth century. Unlike most eastern and midwestern states, Florida was still largely undeveloped in the early twentieth century (Knight et al. 2011; Volk et al. 2017). Across the southeastern US, only the southern third of the Florida Peninsula failed to reach a threshold of two people per square mile by 1890, whereas other portions of the region, especially Virginia and the Carolinas, had reached that threshold by 1770 (Frost 2006). The inaccessibility of much of South Florida explains why a remnant population of puma, the Florida panther (*Puma concolor*), was able to persist here while being exterminated elsewhere in the eastern United States. Amazingly, an analysis of the Florida Cooperative Land Cover Map (CLC 2023) shows 48.7% of Florida's land area remains in natural condition—and this does not include seminatural lands such as rangelands and commercial timberlands. But that percentage is rapidly diminishing.

What the mixed blessing of being a hotspot of irreplaceability and vulnerability means is that land in Florida must be protected at a rate that exceeds its destruction. It also means we must prioritize land conservation wisely so that the most biodiverse places are protected before they are lost forever. Fortunately, Florida has long had a robust conservation prioritization and acquisition process (Henderson 2022; see chapter 3), a large collection of land trusts enabling private land conservation (see chapter 9), a vibrant land-use planning apparatus (see chapter 4), and a rapidly growing network of wildlife crossing structures to mitigate the fragmenting effects of roads (see chapter 8), which collectively have helped reduce the deleterious impacts of growth while protecting many of the most biologically important sites. For a state that has so much to lose if growth continues unabated, we are encouraged that conservationists and political leaders from across the political spectrum have been taking large landscape conservation seriously.

Balancing Conservation of Large, Connected Landscapes and Small, Isolated Sites

As two of the ecologists who, along with our mentor Larry Harris, most adamantly advocated large landscape conservation in Florida—replete with big core areas and corridors for panthers, bears, and other fragmentation-sensitive species—we as editors of this book must humbly acknowledge the limitations of the landscape approach as it is usually pursued. The Florida Ecological Greenways Network (FEGN) and the Florida Wildlife Corridor are essential elements of, and a foundation for, a statewide conservation strategy. However, not all biologists and conservationists are enthusiastic about the Florida Wildlife Corridor or landscape conservation planning generally. Specifically, some of our botanist colleagues are concerned that conservation of large tracts of land—in particular, cattle ranches—of marginal habitat quality through conservation easements detracts from efforts to protect smaller, often isolated sites with higher habitat quality and greater concentrations of rare and endemic species.

In the book *Forgotten Grasslands of the South* (Noss 2012), one of us wrote about the uneasiness he sometimes feels about helping initiate the movement toward regional landscape conservation in North America:

> Beginning in the early 1980s I published a series of papers arguing that conservationists have focused too narrowly on saving small sites with rare species . . . Conservationists, I argued, should emphasize the protection and restoration of large landscapes and "functional mosaics" of communities, which compose the characteristic natural vegetation of a biogeographic region, rather than concentrating on quirky "museum pieces" that are too small to contain viable populations of many species and are difficult to protect and manage. Slowly but surely, my ideas caught on all too well. TNC [The Nature Conservancy] and many other conservation groups jumped on the landscape bandwagon and now have come perilously close to abandoning the protection of inherently small communities and sites with concentrations of endemics and other imperiled species.

Despite our status as outspoken advocates for large landscape conservation, we believe there must be no zero-sum game between large landscape conservation and the traditional focus of the natural areas movement on small sites holding rare species and high-quality examples of natural communities. Both approaches are desperately needed. Across the United States, the average size of protected areas is only around 2,000 ha and the median is 20 ha, yet these sites contribute immensely to conservation (Baldwin and Fouch 2018). It cannot be denied that "collectively, the small, weird sites contain a substantial fraction of a region's biodiversity and often the majority of its truly distinctive and endemic taxa. Their conservation value is complementary to that of large, functional landscapes" (Noss 2012).

Over recent years, many studies have demonstrated the conservation values of small protected areas. Especially in landscapes heavily modified by humans, strategically located small reserves play several key roles that complement large areas, including (1) protecting habitats for range-limited or relic species; (2) conserving threatened or sensitive habitats in altered ecosystems; (3) conserving habitats critical to the life histories of particular species, such as wading bird rookeries or raptor nest sites; and (4) enhancing landscape connectivity by providing stepping-stones of suitable habitat for dispersing individuals (Dudley et al. 2024). They also provide cultural values such as opportunities for people, especially urban dwellers, to connect with wild, natural, or seminatural areas close to home rather than having to travel to larger areas, which are usually more remote.

Moreover, most studies show that a constellation of small reserves well distributed across a region will protect more species—especially of plants—than fewer large reserves of equivalent total area (Armsworth et al. 2018; Wintle et al. 2018; Volenec and Dobson 2020). This finding can be explained by patterns of species distribution. The change in species composition across a region or from site to site, known as beta diversity (Whittaker 1960), can form two distinct patterns (though sometimes they are mixed): (1) nestedness, where large sites contain the most species and smaller sites contain depauperate subsets of these species without adding any new species absent from larger sites; and (2) turnover, where large sites may still contain the most species, but smaller sites are generally not subsets but rather contain unique species not found in the larger sites.

Turnover reflects beta diversity attributable to species replacement from site to site, whereas nestedness indicates species loss or gain, with smaller sites exhibiting species loss (Baselga 2010). Conservation biologists strongly influenced by island biogeographic theory (see chapter 2) assumed nestedness was the predominant pattern, but many empirical studies show turnover is more common. A meta-analysis of nestedness and turnover components of beta diversity showed differences related to latitude, trophic position (level in the food chain), and other factors. In general, consumers (especially carnivores) show nested communities, whereas producers (plants) show mostly turnover (Soininen et al. 2018). In line with this generalization, a study of plant-species distributions in longleaf pine (*Pinus palustris*) savannas of the southeastern Coastal Plain—the dominant natural communities of Florida—found beta diversity patterns are turnover dominated (Freeman et al. 2019). It is not surprising, then, that botanists tend to favor protection of many small reserves, even if isolated, whereas zoologists or wildlife ecologists tend to promote large, connected reserves. Obviously (to us), we need both. We are reminded that, nearly four decades ago, two biologists who had been on opposite sides of the debate concerning small versus large reserves published a paper arguing that both small and large reserves are necessary—and as many as possible (Soulé and Simberloff 1986).

Nevertheless, most studies extolling the virtues of small reserves fail to mention their shortcomings, such as failure to provide sufficient habitat to area-sensitive species, exposure to physical and biological edge effects, frequently high abundance of nonnative species, higher management costs per unit area, and lower probability of ecological resilience. These problems underscore the need for many reserves, both large and small, and more so, a plan for functionally integrated and protected public and private lands that work together to conserve biodiversity and ecosystem services—that is, a regional landscape approach!

We acknowledge that landscape conservation planning in Florida to date has paid too little attention to small reserves and their many values. When people think of regional or landscape conservation planning, maps with big core areas and broad corridors—the Florida Wildlife Corridor and its equivalents in other regions—come to mind, not tiny patches of pine rockland within the Miami-Dade Metropolitan Area or minuscule upland glades in the Panhandle (Noss 2012). We at the University of Florida's Center for Landscape Conservation Planning are currently putting more effort, working with the Florida Natural Areas Inventory and other partners, toward ensuring that small natural areas and local corridors that are important for their biodiversity values, ecosystem services, or cultural values are a significant part of Florida's conservation design. We are developing new spatial models to identify regional to local wildlife and riparian corridors to benefit local conservation planning in areas outside the FEGN and Florida Wildlife Corridor. We are conducting urban green infrastructure modeling and planning that identifies smaller sites with multiple conservation values as well as buffer zone planning that could lead to functional transitions from urban to wildland where those opportunities still exist. We are also working to identify biodiversity and ecosystem service elements that are well represented in the FEGN / Florida Wildlife Corridor and those that are not to highlight the importance of conservation priorities across areas of different sizes and regions.

Most of the Florida coastline was left out of large-scale corridor designs from the earliest statewide corridor maps (Noss 1985, 1987) to the modern FEGN (Hoctor et al. 2000) and Florida Wildlife Corridor (see chapter 7) except in the few regions—such as the Florida Big Bend, Everglades, and portions of the Panhandle and extreme Northeast Florida—where large interior landscapes are still functionally connected to coastal ecosystems. Large, connected landscapes no longer exist along the remainder of Florida's coastline. We are now identifying smaller sites and narrower corridors that fulfill critical functions such as enabling species migration away from the coast to escape sea-level rise, including efforts with the Florida Department of Environmental Protection to identify opportunities to facilitate resilience of the watersheds of Florida's Aquatic Preserves and other important coastal ecosystems. There is also a clear need to identify habitat mosaics across watersheds, coastal areas, and estuaries, which are essential to the life histories of many aquatic species (see chapter 6), and to identify mosaics of ephemeral and other isolated wetlands within upland matrixes that are needed by many wetland-breeding amphibians and other species. We believe that a regional landscape approach to conservation should examine the entire landscape and the potential contribution of sites of all sizes and contexts to conservation goals.

Finally, while this book paints an optimistic picture of conservation in Florida—and that picture is mostly accurate—conservationists here and elsewhere should not rest on their laurels. There is no guarantee that the pace of land protection will keep up with the relentless growth of the human population and all the land destruction that accompanies it. Florida conservationists and those in other states and regions must continue to make the case for significant and sufficient land (and water) conservation funding. While the current governor and legislature are largely supportive of land conservation, as almost all administrations of both parties have been for decades, it is not assured that future administrations will be supportive.

Conservationists also need to be continually vigilant in keeping protected lands protected. In the conservation field we champion the concept of "in perpetuity" protection, but the reality is that there will always be challenges to the status of public and private conservation lands as long as land-use conversion pressure is high and politics remain in play. While work on this book was in progress, the State of Florida approved a new highway to cross the Split Oak Forest Wildlife and Environmental Area, a reserve southeast of Orlando managed by the Florida Fish and Wildlife Conservation Commission. Ironically, this conservation area was established in the 1990s as mitigation for the destruction of habitat for development nearby. The new highway was proposed not to meet current traffic needs but rather to facilitate the development of a huge new city by the Church of Jesus Christ of Latter-day Saints and its developer partners, also approved by the state. As Orange County Commissioner Nicole Wilson, one of the handful of local politicians who opposed the Split Oak expressway, put it, "Nothing is sacred. If our public parks can be hacked away and sliced by destructive toll roads, nothing is safe" (Maxwell 2024). A similar decision was made to construct a new road through Serenova Preserve in Pasco County, which (again ironically) had been protected with mitigation funding from the environmental impacts of the nearby Suncoast Parkway.

Split Oak and Serenova demonstrate that despite relatively strong protections for state-owned conservation lands in Florida, there is always pressure for surplusing conservation areas and allowing incompatible uses. As we were writing this chapter, the Florida Department of Environmental Protection, with backing from the governor, released plans for massive tourist developments on state parks across the state, including golf courses, pickleball courts, and hotels built in imperiled natural communities such as Florida scrub and maritime hammock (Payne 2024), which mirrors similar efforts for incompatible development within state parks attempted a decade ago. In both cases, the reaction from state park–loving Floridians, including both Republican and Democratic legislators, was swift and vehement, which made clear the popularity of state conservation lands is a significant impediment to development, though perhaps not unassailable when political pressure is high enough. Even conservation easements, legally required to be protected in perpetuity, are not entirely safe from development in Florida. Several regulatory conservation easements have been "released," meaning opened for development, prompting legal scholars to call for a revision of Florida's conservation easement–enabling statute (McLaughlin 2024; see chapter 9). Both public and private conservation lands need strong protection and reliable public support to ensure they remain managed for purposes compatible with the state's conservation goals.

Only an engaged citizenry active in the democratic process might succeed in protecting conservation areas—in Florida and elsewhere—truly in perpetuity. As usual, Aldo Leopold said it best: "A militant minority of wilderness-minded citizens must be on watch throughout the nation and vigilantly available for action" (Leopold 1949). Regional landscape conservation planning can help ensure that conservation action is efficiently directed to the sites that are most irreplaceable and vulnerable. A primary goal of this book is to provide engaged citizens with accessible explanations of the importance of landscape-scale conservation and the opportunities and obstacles for implementing ambitious conservation plans.

Literature Cited

Armsworth, P. R., H. B. Jackson, S.-H. Cho, M. Clark, J. E. Fargione, G. D. Iacona, T. Kim, E. R. Larson, T. Minney, and N. A. Sutton. 2018. Is conservation right to go big? Protected area size and conservation return-on-investment. Biological Conservation 225:229–36.

Baldwin, R. F., and N. T. Fouch. 2018. Understanding the biodiversity contributions of small protected areas presents many challenges. Land 7:123. https://doi.org/10.3390/land7040123.

Baselga, A. 2010. Partitioning the turnover and nestedness components of beta diversity. Global Ecology and Biogeography 19:134–143.

Biodiversity of North America Project (BONAP). Undated. Density gradient of native vascular plant families per 10,000 km^2. http://bonap.org/2015_SpecialtyMaps/Density%20Gradient%202015/original/orano3_2014_Families.png. Accessed August 2, 2024.

Ceballos, G., and P. R. Ehrlich. 2023. Mutilation of the tree of life via mass extinction of animal genera. Proceedings of the National Academy of Sciences 120(39):e2306987120.

Connell, J. H. 1978. Diversity in tropical rain forests and coral reefs. Science 199:1302–1310.

Cooperative Land Cover (CLC). November 2023. Florida Cooperative Land Cover Map. https://myfwc.com/research/gis/wildlife/cooperative-land-cover/. Accessed September 26, 2024.

Cowling, R. M., R. L. Pressey, M. Rouget, and A. T. Lombard. 2003. A conservation plan for a global biodiversity hotspot—the Cape Floristic Region, South Africa. Biological Conservation 112:191–216.

Critical Ecosystem Partnership Fund. Undated. Hotspots: North American Coastal Plain. https://www.cepf.net/our-work/biodiversity-hotspots/north-american-coastal-plain. Accessed July 30, 2024.

Dobzhansky, T. 1951. Genetics and the Origin of Species. Columbia University Press, New York.

Dudley, N., H. L. Timmins, S. Stolton, and J. E. M. Watson. 2024. Effectively incorporating small reserves into national systems of protected and conserved areas. Diversity 16:216. https://doi.org/10.3390/d16040216.

Fenneman, N. M. 1938. Physiography of Eastern United States. McGraw-Hill, New York.

Freeman, J. E., L. N. Kobziar, E. H. Leone, and K. Williges. 2019. Drivers of plant functional group richness and beta diversity in fire-dependent pine savannas. Diversity and Distributions. https://doi.org/10.1111/ddi.12926.

Frost, C. C. 2006. History and future of the longleaf pine ecosystem. Pages 9–48 in S. Jose, E. Jokela, and D. Miller, editors. The Longleaf Pine Ecosystem: Ecology, Silviculture, and Restoration. Springer-Verlag, New York.

Gentry, A. H. 1986. Endemism in tropical versus temperate plant communities. Pages 153–181 in M. E. Soulé, editor. Conservation Biology: The Science of Scarcity and Diversity. Sinauer, Sunderland, Massachusetts.

Germain-Aubrey, C. C., P. S. Soltis, K. M. Neubig, T. Thurston, D. E. Soltis, and M. A. Gitzendanner. 2014. Using comparative biogeography to retrace the origins of an ecosystem: The case of four plants endemic to the Central Florida scrub. International Journal of Plant Sciences 175:418–431.

Grime, J. P., and S. Pierce. 2012. The Evolutionary Strategies That Shape Ecosystems. Wiley-Blackwell, West Sussex, UK.

Hamilton, H., R. L. Smyth, B. E. Young, T. G. Howard, C. Tracey, S. Breyer, D. R. Cameron, A. Chazal, A. K. Conley, C. Frye, and C. Schloss. 2022. Increasing taxonomic diversity and spatial resolution clarifies opportunities for protecting imperiled species in the US. Ecological Applications 2022:e2534. https://doi.org/10.1002/eap.2534.

Harrison, S., and R. Noss. 2017. Endemism hotspots are linked to stable climatic refugia. Annals of Botany 119:207–214.

Henderson, C. 2022. Forces of Nature: A History of Florida Land Conservation. University Press of Florida, Gainesville.

Huston, M. 1979. A general hypothesis of species diversity. American Naturalist 113:83–101.

James, C. W. 1961. Endemism in Florida. Brittonia 13:225–244.

Knight, A. T., R. M. Cowling, M. Rouget, A. Balmford, A. T. Lombard, and B. M. Campbell. 2008. Knowing but not doing: Selecting priority conservation areas and the research-implementation gap. Conservation Biology 22:610–617.

Knight, G. R., J. B. Oetting, and L. Cross. 2011. Atlas of Florida's Natural Heritage: Biodiversity, Landscapes, Stewardship, and Opportunities. Florida State University, Tallahassee.

Kremen, C., A. Cameron, A. Moilanen, S. J. Phillips, C. D. Thomas, H. Beentje, et al. 2008. Aligning conservation priorities across taxa in Madagascar with high-resolution planning tools. Science 320:222–226.

Lande, R. 1993. Risks of population extinction from demographic and environmental stochasticity and random catastrophes. American Naturalist 142:911–927.

Mann, C. 2005. 1491: New Revelations of the Americas before Columbus. Knopf, New York.
Margules, C. R., and R. L. Pressey. 2000. Systematic conservation planning. Nature 405:243–253.
Maxwell, S. 2024. The Split Oak betrayal: Will Florida ever change? Orlando Sentinel, July 12, 2024. https://www.orlandosentinel.com/2024/07/11/split-oak-betrayal-maxwell/. Accessed August 24, 2024.
Mayr, E. 1942. Systematics and the Origin of Species. Columbia University Press, New York.
McLaughlin, N. A. 2024. Keeping the perpetual in Florida's conservation easements. FIU Law Review 18:347–383. https://dx.doi.org/10.25148/lawrev.18.2.9. Accessed August 24, 2024.
Meine, C., M. Soulé, and R. F. Noss. 2006. "A mission-driven discipline": The growth of conservation biology. Conservation Biology 20:631–651.
Mishler, B. D., R. Guralnick, P. S. Soltis, S. A. Smith, D. E. Soltis, N. Barve, et al. 2020. Spatial phylogenetics of the North American flora. Journal of Systematics and Evolution 58:393405.
Mittermeier, R. A., W. R. Turner, F. W. Larsen, T. M. Brooks, and C. Gascon. 2011. Global biodiversity conservation: The critical role of hotspots. Pages 3–22 in F. E. Zachos and J. C. Habel, editors. Biodiversity Hotspots: Distribution and Protection of Conservation Priority Areas. Springer, Heidelberg.
Mittermeier, R. A., P. P. van Dijk, A. G. J. Rhodin, and S. D. Nash. 2015. Turtle hotspots: An analysis of the occurrence of tortoise and freshwater turtles in biodiversity hotspots, high-biodiversity wilderness areas, and turtle priority areas. Chelonian Conservation and Biology 14:2–10.
Morgan, G. S., and S. D. Emslie. 2010. Tropical and western influences in vertebrate faunas from the Pliocene and Pleistocene of Florida. Quaternary International 217:143–158.
Myers, J. A., and K. E. Harms. 2011. Seed arrival and ecological filters interact to assemble high-diversity plant communities. Ecology 92:676–686.
Myers, N., R. A. Mittermeier, C. G. Mittermeier, G. A. da Fonseca, and J. Kent. 2000. Biodiversity hotspots for conservation priorities. Nature 403:853–858.
NatureServe Network. 2024. The Map of Biodiversity Importance. Arlington, Virginia.
Noss, R. F. 2012. Forgotten Grasslands of the South: Natural History and Conservation. Island Press, Washington, DC.
Noss, R. F. 2018. Fire Ecology of Florida and the Southeastern Coastal Plain. University Press of Florida, Gainesville.
Noss, R. F., C. Carroll, K. Vance-Borland, and G. Wuerthner. 2002. A multicriteria assessment of the irreplaceability and vulnerability of sites in the Greater Yellowstone Ecosystem. Conservation Biology 16:895–908.
Noss, R. F., M. A. O'Connell, and D. D. Murphy. 1997. The Science of Conservation

Planning: Habitat Conservation under the Endangered Species Act. Island Press, Washington, DC.

Noss, R. F., and R. L. Peters. 1995. Endangered Ecosystems of the United States: A Status Report and Plan for Action. Defenders of Wildlife, Washington, DC.

Noss, R. F., W. J. Platt, B. A. Sorrie, A. S. Weakley, D. B. Means, J. Costanza, and R. K. Peet. 2015. How global biodiversity hotspots may go unrecognized: Lessons from the North American Coastal Plain. Diversity and Distributions 21:236–244. https://doi.org/10.1111/ddi.12278.

Olson, D. M., and E. Dinerstein. 1998. The Global 200: A representation approach to conserving the Earth's most biologically valuable ecoregions. Conservation Biology 12:502–515.

Olson, D. M., E. Dinerstein, E. D. Wikramanayake, N. D. Burgess, G. V. N. Powell, E. C. Underwood, et al. 2001. Terrestrial ecoregions of the world: A new map of life on Earth. BioScience 51:933–938.

Payne, K. 2024. Floridians balk at DeSantis administration plan to build golf courses at state parks. Associated Press, August 23, 2024. https://apnews.com/article/florida-desantis-state-parks-golf-pickleball-development-f6f1e129cdd316207eeabc90edf2aad6. Accessed August 24, 2024.

Pressey, R. L., I. R. Johnson, and P. D. Wilson. 1994. Shades of irreplaceability: Towards a measure of the contribution of sites to a reservation goal. Biodiversity and Conservation 3:242–262.

Pyne, S. J. 2016. Florida: A Fire Survey. University of Arizona Press, Tucson.

Sarasota Herald-Tribune. 2024. Florida population tops 23 million for first time. Where are people moving from? https://www.heraldtribune.com/story/news/2024/07/23/florida-population-growth-new-york-california/74508541007/. Accessed August 19, 2024.

Soininen, J., J. Heino, and J. Wang. 2018. A meta-analysis of nestedness and turnover components of beta diversity across organisms and ecosystems. Global Ecology and Biogeography 27:96–109.

Sorrie, B. A., and A. S. Weakley. 2001. Coastal Plain plant endemics: Phytogeographic patterns. Castanea 66:50–82.

Soulé, M. E. 1985. What is conservation biology? BioScience 35:727–734.

Soulé, M. E., and D. Simberloff. 1986. What do genetics and ecology tell us about the design of nature reserves? Biological Conservation 35:19–40.

Stebbins G. L., and J. Major. 1965. Endemism and speciation in the California flora. Ecological Monographs 35:1–35.

Takhtajan, A. 1986. Floristic Regions of the World. University of California Press, Berkeley.

United States Census Bureau. 2022. New Florida estimates show nation's third-largest state reaching historic milestone. December 22, 2022. https://www.census.gov

/library/stories/2022/12/florida-fastest-growing-state.html. Accessed August 20, 2024.

Volenec, Z. M., and A. P. Dobson. 2020. Conservation value of small reserves. Conservation Biology 34:66–79.

Volk, M. I., T. S. Hoctor, B. B. Nettles, R. Hilsenbeck, F. E. Putz, and J. Oetting. 2017. Florida land use and land cover change in the past 100 years. Pages 51–82 in E. P. Chassignet, J. W. Jones, V. Misra, and J. Obeysekera, editors. Florida's Climate: Changes, Variations, and Impacts. Gainesville: Florida Climate Institute.

Webb, S. D. 1977. A history of savanna vertebrates in the New World. Part I: North America. Annual Review of Ecology and Systematics 8:355–380.

Whittaker, R. H. 1960. Vegetation of the Siskiyou Mountains, Oregon and California. Ecological Monographs 30:280–338.

Wintle, B. A., H. Kujala, A. Whitehead, A. Cameron, S. Veloz, A. Kukkala, A. Moilanen, A. Gordon, P. E. Lentinia, N. C. R. Cadenhead, and S. A. Bekessy. 2018. Global synthesis of conservation studies reveals the importance of small habitat patches for biodiversity. Proceedings of the National Academy of Sciences 116:909–914. https://www.pnas.org/doi/full/10.1073/pnas.1813051115.

2

The Influences, Precursors, and Early History of Regional Landscape Conservation Planning

Reed F. Noss and Thomas S. Hoctor

Introduction

The mid-1970s through the 1990s were heady times to be an ecologist interested in conservation. The sciences of conservation biology, landscape ecology, and restoration ecology emerged and blossomed during these years; ecological theory began to be applied to real-world conservation; technologies such as geographic information systems (GIS) and niche and population modeling appeared; and a new grassroots conservation movement grew out of discontent with mainstream environmentalism and its perceived deficiency of vision. Each of these developments had precursors much earlier, but they did not come together cohesively until the final three decades of the twentieth century. This was a time when, for a variety of reasons, the need for a landscape perspective in conservation became increasingly evident (Trombulak and Baldwin 2010).

As frequently observed in history, ideas and plans that were deemed radical and hopelessly idealistic when first proposed eventually spread from the fringe toward the center of thinking in conservation. Florida was a critical geography for this convergence of concepts and approaches that ultimately led to the science of regional landscape conservation planning as we know it today. The long history of applied conservation science in Florida culminated in the Florida Ecological

Greenways Network (FEGN) and the Florida Wildlife Corridor, established as a policy priority in Florida and enshrined in state law. These projects continue to inspire other ambitious conservation plans around the world.

In this chapter we review the early history of regional landscape conservation planning up to the origin of the FEGN, including the key people, events, and ecological theories that facilitated the development of landscape conservation (for a review of the history of the FEGN and modern Florida Wildlife Corridor, see chapter 7). The summaries that follow are mere vignettes of much richer and deeper stories. Our focus is not exclusively on Florida; we hope to show that the Florida model of large landscape conservation was a key element in a global trend toward bigger and bolder thinking in conservation (Noss et al. 2012).

Influences and Precursors

ECOSYSTEM REPRESENTATION AND CONSERVATION

At its heart, regional landscape conservation planning is about identifying, protecting, and restoring intact ecosystems and their constituent species and processes. Landscapes are large, heterogeneous ecosystems, and the ecosystem is a fundamental unit of study in the science of ecology. British ecologist Arthur Tansley usually gets credit for coining the term "ecosystem" because he was the first to use the word in print (Tansley 1935). Earlier, other ecologists such as Stephen Forbes (a primary founder of the science of ecology in North America; Croker 2001) had developed similar concepts of systems with interacting biotic (living) and abiotic (nonliving) components (e.g., Forbes 1887).

Tansley characterized the ecosystem as "the whole system (in the sense of physics), including not only the organism-complex, but also the whole complex of physical factors forming what we call the environment of the biome—the habitat factors in the widest sense" (Tansley 1935). Tansley declared ecosystems "the basic units of nature on the face of the earth," a view held by many ecologists today. And like modern ecologists, Tansley recognized that ecosystems "are of the most various kinds and sizes" (Tansley 1935).

Victor E. Shelford, a pioneering ecologist who spent most of his career at the University of Illinois, was committed to representing all major ecosystem types in North America within large conservation areas. His extensive knowledge of the continent's natural history positioned him uniquely to develop the world's first continental-scale ecosystem conservation strategy—before the term "ecosystem" had been coined. In 1915 Shelford cofounded the Ecological Society of America (ESA) and, beginning in 1916, served as its first president. Critical to our story, in 1917 Shelford initiated ecosystem conservation research and advocacy within the ESA with the establishment of the Committee on the Preservation of Natural Conditions, which he chaired. A significant early product of this committee was the 761-page *Naturalist's Guide to the Americas* (Shelford 1926), published with Shelford as primary editor and with assisting editors including eminent ecologists E. Lucy Braun, Forrest Shreve, and Lee R. Dice. This massive tome is primarily a compendium of the regions, biomes, and vegetation types of North America and the Neotropics north of the equator, along with descriptions of existing parks as well as unprotected natural areas representative of each ecosystem and worthy of protection. In 1931 a separate committee devoted to research, the Committee for the Study of Plant and Animal Communities, was formed to advise the Preservation Committee (Croker 1991).

The explicit goal of Shelford's committees was to establish a continent-wide network of protected areas to represent each major ecosystem or biome. The reasons given for the need of such a network emphasized scientific research but also included recreational, aesthetic, and other purposes. Natural areas were described as "living museums" for research and education. As a start, the Preservation Committee called for "an undisturbed area in every national park and public forest." This goal was quickly expanded into a more ambitious resolution to establish "a nature sanctuary with its original wild animals for each biotic formation," which was accepted by the ESA Governing Board in 1931 (Croker 1991).

A zoologist by training, Shelford was not content with conserving representative examples of vegetation types devoid of their characteristic animals. He advocated protection of wild ecosystems with their complete native faunas and intact food webs. In a 1933 paper in *Ecology*, he emphasized that sanctuaries must be large enough to sustain populations of the most area-demanding and often controversial species: "The animals requiring first and most careful consideration are the carnivores, likely to be unpopular with the agricultural . . . interests outside the park or forest. The home range of these animals must be considered . . . A second group demanding careful study is the migratory herbivores" (Shelford 1933a). Shelford wanted sanctuaries large and wild enough to be essentially autonomous and self-managing. In concluding his *Ecology* paper, he suggested that "the experiment of letting areas essentially alone, so successful in a few of our parks, is worthy of repetition."

Meanwhile, another visionary, US National Park Service biologist George

Melendez Wright, promoted a similar theme of maintaining national parks in a wild and unimpaired condition with all their native wildlife thriving. Such ideas were revolutionary at a time in national park history when park managers routinely killed predators, and the primary goal of parks was to provide scenic views to tourists. Wright is credited with initiating science-based management within the national park system. His groundbreaking work was cut short by his death in an automobile accident in 1936 at the age of 31 (Duncan 2009).

Continental-scale maps produced by Shelford's Preservation Committee during the 1930s displayed core areas centered largely on existing national parks and surrounded by wide buffer zones but also included new proposed sanctuaries for ecosystems not represented in current parks, including large reserves in central Idaho, Arkansas, Oklahoma, and Nebraska among other regions. Shelford was particularly interested in protecting and restoring grasslands, which he recognized were especially depleted and endangered in North America. As early as 1931 he called for a grasslands national park on the order of one million acres (Croker 1991). Writing in *Science*, Shelford explained that nature sanctuaries must sustain naturally dynamic ecosystems and populations: "In nature sanctuaries the natural fluctuations of organisms are allowed free play and serve among other things to show what natural fluctuations in abundance are like" (Shelford 1933b).

By 1943 the leaders of the ESA had grown uncomfortable with the advocacy activities of Shelford's committees, believing that ESA should stick to science and not get involved in public affairs except in an advisory capacity (Croker 1991). Shelford insisted that it was the duty of ecologists to apply their knowledge for the common good, writing that "human society, which supports research, will hold scientific men and the societies which they constitute responsible for failure to urge the application of their knowledge directly and simply whenever it is in the interest of society to do so" (Shelford 1944).

Shelford lost his battle with the conservative governing board of the ESA. His committees were disbanded by board decree. Disappointed, angry, but undeterred, in 1946 Shelford and colleagues organized an independent group, the Ecologists' Union, to continue the work of the former ESA committees (Croker 1991). A report credited jointly to the Ecologists' Union and the ESA's Committee on the Study of Plant and Animal Communities was published in *The Living Wilderness* (the journal of the Wilderness Society) in the winter of 1950–1951. This report documented that no protected areas large enough to contain all native animal species in self-maintaining populations existed for deciduous forests, prairies, or lower elevations of the Rocky Mountains in the US and Canada. Nevertheless, opportunities to create such sanctuaries remained in some southern swamps, deserts, higher elevations in western mountains, boreal forests, and tundra (Kendeigh et al. 1950–1951). In September of 1950, the Ecologists' Union was reorganized and changed its name to The Nature Conservancy. This initially small organization was led by Stanley Cain (president), George B. Fell (vice president), and Joseph Hickey (secretary-treasurer) (Croker 1991). Before long, The Nature Conservancy (TNC) grew into the largest and most successful land conservation organization in the world (Pearson 2017).

The stated purpose of TNC was "to preserve or aid in the preservation of all types of wild nature including natural areas, features, objects, flora and fauna and biotic communities" (Pearson 2017). For a while, TNC pursued Shelford's mission of protecting examples of ecosystem types through its "coarse filter" ecosystem representation strategy, which was paired with the complementary "fine filter" of locating and protecting populations of rare species. A major argument for the coarse filter was that there are too many species to address individually, and many of them are poorly known. In addition, the first stated goal of the US Endangered Species Act of 1973 is "to provide a means whereby the ecosystems upon which endangered or threatened species depend may be conserved." In 1974 TNC's vice president for science, Robert E. Jenkins, developed the basic natural heritage methodology and established the first state natural heritage program in South Carolina (Jenkins 1978, 1985).

WILDERNESS RECOVERY AND REWILDING

The original Florida Wildlife Corridor in the 1980s (Noss 1985, 1987) was devoted to rewilding at least half of the state through recovery of large carnivores and restoration of natural disturbance regimes and other ecological processes. The Florida panther (*Puma concolor*) and Florida black bear (*Ursus americanus floridanus*), as the most wide-ranging terrestrial animals, were the primary focal species for planning. They remain so today for the design and implementation of the FEGN and the Florida Wildlife Corridor. After its early development in Florida and other regions of North America (see chapter 10), rewilding became an international campaign (Carver et al. 2021).

The origin of the rewilding idea is deeper than the Florida experience. The concept emerged from two distinct but ultimately converging streams—the science of ecology and the grassroots wilderness preservation movement in twentieth-century North America. As noted previously, the representative protected areas advocated by Victor Shelford and his colleagues were intended to support intact food webs replete with apex predators and to function naturally without direct human intervention. Many biomes, however, had already been so heavily modified by human activities that no large, representative sites remained. For such biomes, Shelford envisioned restoration of wilderness conditions. Thus, Shelford advocated rewilding, albeit not by that name. In a memo to his study committee in 1931, Shelford asserted, "There needs to be restoration of the primitive as well as protection of the primitive" (Croker 1991). At around the same time, George Wright advocated restoration of wilderness character within national parks (Sellars 2000).

Restoration of primitive conditions and ecological integrity on a broad scale, as advocated by Shelford and Wright, did not get much attention from environmentalists in North America until a half-century later. In November 1980, in the first issue of the *Earth First!* newsletter, an editorial proclaimed the principles of the new movement. The first principle stated, "Protection of some of our remaining wilderness is not enough. Protection of all of it is not enough . . . we also believe that in order to preserve the ecological integrity of our country, it is time to recreate wilderness: identify key areas, close roads, remove developments, and reintroduce extirpated wildlife" (Anonymous 1980). The following year, an article in *The Progressive* expanded on this idea: "Pick an area for each of our major ecosystems and recreate the American wilderness—not in little pieces of a thousand acres but in chunks of a million or ten million. . . . It is not enough any longer to say no more dams on our wild rivers. We must begin tearing down some dams already built" (Foreman 1981). Dam removal has since become a widely accepted approach to watershed restoration (Foley et al. 2017). Earth First! advocated reduced emphasis on the scenic and recreational values of wilderness (which had dominated the American wilderness movement) and, echoing Shelford, a shift in focus to "ecological wilderness" with complete food webs and natural disturbance regimes.

Over the succeeding decade of the 1980s, several proposals for re-creating wilderness were published in *Earth First!* and elsewhere. The term "wilderness recovery areas" was introduced in an editorial in March 1983, which proposed that "these areas will be managed to assure that they will revert to a wilderness condition" (Wolke 1983). Specific wilderness recovery networks with accompanying maps were proposed for the Ohio River Valley (Noss 1983), Florida (Noss 1985), the Appalachian Mountains (Sayen 1987), and elsewhere (see chapter 10).

The early wilderness recovery proposals were science-informed but not highly rigorous. They followed simple principles of reserve design (Diamond 1975; Wilson and Willis 1975), especially "bigger is better" and "connected is better than isolated." As these times were pre–GIS technology, maps of proposed reserve networks were hand-drawn and crude by today's standards. Wilderness recovery plans invariably proposed recovery and often reintroduction of large carnivores and herbivores, beaver, and other ecologically pivotal species that had been extirpated or severely diminished, as well as restoration of fundamental processes such as fires and floods (Noss 1986). Modern regional rewilding proposals (e.g., Ripple et al. 2022) have the same ingredients.

In 1991, several former editors and writers for *Earth First!* along with some academic ecologists, environmental leaders, and donors founded the Wildlands Project (originally called the North American Wilderness Recovery Strategy) and its associated journal, *Wild Earth* (published 1991–2004). These wilderness traditionalists—who included Earth First! leader Dave Foreman, prominent conservation philanthropist Doug Tompkins, the Society for Conservation Biology's (SCB's) first president Michael Soulé, and a later SCB president and editor of *Conservation Biology*, Reed Noss—were committed to wilderness, wildness, and biodiversity as the principal values of the ecocentric conservation movement. The Wildlands Project later changed its name to Wildlands Network but continued its mission of designing and promoting regional reserve networks or "wildways" across North America (see chapter 10).

Around the time the Wildlands Project was founded, Foreman began using "rewilding" as a synonym for "wilderness recovery." The term "rewilding" first appeared in print in a *Newsweek* article concerning the desire of radical environmentalists (i.e., Earth First!) to protect or restore wilderness conditions across one-third or more of the North American continent (Foote 1990). That seemed a radical proposal at the time. Today, more than three decades later, we have an internationally accepted goal of protecting 30% of the Earth's land and water by 2030. In late 2022, the Convention on Biological Diversity (United Nations Global Biodiversity Framework) in Montreal agreed to advance "effective conservation and management of at least 30% of the world's lands, inland waters, coastal areas and oceans, with emphasis on areas of particular importance for biodiversity and ecosystem functioning and services" (UNEP 2022); planning is underway for conserving 50% of global lands and waters by 2050.

Again, what was once a fringe idea eventually became mainstream. The maturing sciences of ecology and conservation biology fully supported the need for land and water conservation on a massive scale to conserve biodiversity and functional ecosystems, along with the ecosystem services they provide. Indeed, in 1972 the eminent ecologist brothers Eugene P. Odum at the University of Georgia and Howard T. Odum at the University of Florida developed a systems ecology model for South Florida, focused on long-term sustainability of life-support systems for human society. They determined "a 1:1 ratio of natural to developed environment would provide a basis for an optimum environmental-use program." They followed with a general recommendation: "Until this kind of systems analysis procedure can be refined and becomes a basis for political action, it would be prudent for planners everywhere to strive to preserve 50 percent of the total environment as natural environment" (Odum and Odum 1972). Two decades later, examinations of the "How much is enough?" question from a biodiversity conservation perspective independently arrived at the same estimate (Noss 1992). These ideas ultimately inspired development of the global Nature Needs Half movement (Locke 2013, 2018), popularized by E. O. Wilson in his book *Half-Earth* (Wilson 2016).

Discussion of rewilding in the environmental literature slowly increased during the 1990s, then more rapidly after the first article on the strategy was published (Soulé and Noss 1998). Throughout this period, the emphasis of rewilding was on the protection of large wild core reserves connected by broad habitat corridors to facilitate recovery of large carnivores and other strongly interactive species—the "3 Cs" of cores, corridors, and carnivores (Soulé and Noss 1998). Most proposed rewilding networks are large, in line with the increasing evidence that approximately 50% of an ecoregion needs to be protected or restored to meet well-established conservation goals (Noss and Cooperrider 1994; Noss 1996; Pressey et al. 2003; Noss et al. 2012; Locke 2014; Dinerstein et al. 2017, 2019). Interestingly, 30% of Florida's land area is already within conservation areas, so we have met the interim UNEP 2030 goal. Moreover, the conserved and opportunity areas of the Florida Wildlife Corridor total 49.7% of Florida's land area, whereas the entire FEGN totals 64.6% (see chapter 7), more ambitious than the "radical" wilderness recovery proposals of the 1980s.

As the idea of rewilding spread around the globe from the late 1990s onward, it took on different meanings in different places. Especially in Europe, an emphasis on pristine wilderness was reduced in favor of adapting rewilding strategies to diverse regions and cultures. European conservation planners developed sophisticated proposals for multiuse regional ecological networks and greenways (Jongman and Pungetti 2004). Low-intensity agricultural landscapes are often emphasized in European rewilding plans, along with de-domestication of livestock to mimic the keystone ecological roles of extinct native megaherbivores such as the aurochs (*Bos primigenius*), the direct ancestor of modern cattle (Jepson 2025).

This broadening of perspective is appropriate. Just as there is a continuum of wildness across landscapes, there is also a continuum of rewilding (Noss 2020). Wildness and rewilding are not absolutes; they occur along a broad spectrum of human influence (Carver et al. 2021). We can find wildness and practice rewilding across landscape contexts and spatial scales ranging from vacant urban lots and backyards to agricultural landscapes to vast roadless areas. In Florida this concept includes recognition that low-intensity land uses (often called "working lands") such as cattle ranches and timberlands, though far from pristine, can contribute to large landscape conservation and the viability of fragmentation-sensitive species. However, the original objective of re-creating large wilderness areas with complete food webs—including apex predators—and naturally functioning ecological processes remains as crucial today as ever.

ISLAND BIOGEOGRAPHIC THEORY AND CONCERNS ABOUT HABITAT FRAGMENTATION

A dominant influence on conservation biology in the late twentieth century was the theory of island biogeography (MacArthur and Wilson 1967). This theory predicts that the diversity of species on any island or island-like habitat is a function of the size of the island (which increases diversity through the species-area effect) and the island's isolation (which decreases diversity by reducing colonization). Large islands connected by dispersal to a species source are predicted to have the highest species diversity, whereas small, isolated islands are predicted to have far fewer species. MacArthur and Wilson (1967) used an example of agricultural habitat fragmentation in Wisconsin (Curtis 1956) in the introduction to their book, and ecologists since Curtis have graphically documented the patterns and consequences of fragmentation in terrestrial landscapes (Burgess and Sharpe 1981; Noss et al. 2006; Haddad et al. 2015; Fletcher et al. 2018).

Although island biogeographic theory was often invoked to explain the impacts of habitat fragmentation, the theory is too simplistic to address many issues related to habitat patterns in real landscapes. For example, problems of edge effects, habitat quality within patches, habitat quality or permeability of the landscape matrix, or disruption of ecological processes are ignored in the theory. Often effects of habitat loss are confounded with effects of fragmentation per se (Fahrig 2017). A synthesis of studies showed that whereas patch area and isolation are important factors explaining patch occupancy for some species, they are poor predictors for many others, especially those for which habitat quality varies among patches (Prugh et al. 2008). These authors concluded that improving matrix quality may lead to better conservation outcomes than manipulating the size and configuration of remnant patches (if that were possible). Although the predictions of island biogeographic theory didn't always hold up in real landscapes, the theory had enormous importance by enlightening scientists and conservationists to the influence of habitat area and connectivity (the opposite of isolation) on population viability and species diversity.

By the mid-1970s, several authors, including Diamond (1975) and Wilson and Willis (1975), suggested that island biogeographic theory can be usefully applied to the design of nature reserves in regions being fragmented by human land uses. These authors proposed a series of rules for reserve design, some of which have held up to empirical scrutiny better than others. One rule that failed to stand up was that a single large block of habitat is better than several smaller blocks of equivalent total area. Too many other variables, such as habitat quality and species turnover along environmental gradients, make such a general rule untenable. Nevertheless, two rules that withstood the test of time are (1) all else being equal, large blocks of habitat will contain—and retain over time—more species than small blocks; and (2) patches connected by corridors (or a permeable matrix) are better than disconnected patches (Noss et al. 1997). A comprehensive global study of the impacts of fragmentation on plants and animals confirmed these generalizations, concluding that "fragmented landscapes have lower diversity, at all scales, than do unfragmented landscapes" and that "conservation efforts and policies should seek to prevent fragmentation and habitat alteration and focus on connecting what are often increasingly isolated fragments" (Goncalves-Souza et al. 2025).

CONNECTIVITY AND METAPOPULATIONS

One of the best-validated rules of conservation planning is that reserves or other habitat patches that are connected by habitat corridors or a landscape matrix permeable to movements of organisms are more likely than isolated patches to retain their native species. Of the potential mechanisms that may explain this pattern, a prominent one is that animals with large area requirements may not be able to maintain populations—or sometimes not even an individual home range—in a relatively small patch of habitat. Such animals need to move across landscapes. This need is elevated if a species requires several distinct habitat types to meet its life history needs. For example, Florida black bear habitat preferences in the Ocala National Forest shift from pine flatwoods in winter and spring to oak scrub in summer and fall, likely due to seasonal shifts in food availability, especially abundant acorns in scrub in late summer and fall (Wooding and Hardisky 1994). Many aquatic species, including sport fish, exhibit ontogenetic niche shifts and make seasonal movements across several distinct habitats (see chapter 6).

Not only large animals require connectivity. Game managers since the early twentieth century have noted that small mammals often use vegetated strips of habitat for movement. Wegner and Merriam (1979) showed that birds and small mammals use fencerows more often than open fields to travel between woodlots in an agricultural landscape. Many empirical studies and reviews demonstrated that habitat corridors facilitate movement and contribute to population persistence and species richness of animals as well as plants (Beier and Noss 1998; Bennett 1998; Haddad et al. 2003; Damschen et al. 2006; Gilbert-Norton et al. 2010). Wildlife corridors were incorporated into conservation plans since the 1970s—for example, an international linkage between Italy and Switzerland, created in 1977 to facilitate the seasonal migration of red deer (*Cervus elaphus*), and a system of corridors linking protected areas in Sri Lanka for the benefit of elephants and other wide-ranging forest animals (Harris and Scheck 1991).

Progress in metapopulation theory (Levins 1970; Hanski 1998, 1999), which is related to island biogeography, bolstered arguments for maintaining or restoring connectivity among habitat patches. A metapopulation is a system of local populations connected by occasional migration (dispersal). Dispersal provides for gene flow and demographic and genetic rescue effects—that is, immigration can rescue a small local population from decline or from genetic deterioration stemming from inbreeding or genetic drift (Brown and Kodric-Brown 1977; Sutherland et al. 2012). If there is too much dispersal and subsequent synchrony of population dynamics, however, the result is a single spatially subdivided population, not a metapopulation. Synchrony of population processes suggests that all subpopulations increase or decline at approximately the same time, which could lead to near-simultaneous extinction of all local populations (Sutherland et al. 2012). On the other hand, what appears to be a metapopulation might be a set of independent populations separated by dispersal barriers in the landscape. Populations inhabiting remnant patches within fragmented landscapes may be completely isolated by a hostile matrix and will not demonstrate the equilibrium colonization-extinction dynamics expected for classic metapopulations (Harrison 1991; Hanski and Simberloff 1997; With 2004).

An important lesson for conservation is that a well-connected network of reserves can be a whole greater than the sum of its parts, in the sense that the network might maintain viable populations or metapopulations of wide-ranging species even though each individual reserve is too small to protect even a single population (Noss and Harris 1986). Although there is good reason to be skeptical of claims that a habitat corridor can override the effects of overall habitat loss on species persistence, studies of functional connectivity demonstrate that well-designed corridors do enhance persistence (e.g., Damschen et al. 2006).

Connectivity is usually most needed by area-sensitive species—those that require some relatively large minimum area to sustain a population or metapopulation. The most area-sensitive terrestrial species are typically large carnivores or migratory ungulates, whose populations operate on regional scales (Noss and Cooperrider 1994). Large carnivores and other apex predators often play "keystone" roles in ecosystems, including regulation of the populations or distribution of their prey, thereby reducing negative impacts of herbivory on vegetation and on species (such as shrub-nesting songbirds) dependent on a particular vegetation structure (Estes et al. 2011). In Florida, a large carnivore, the Florida panther, is a primary focal species for statewide connectivity and rewilding designs.

Among the other effects of the "trophic cascades" that occur when populations of large carnivores are extirpated are increases in populations of medium-sized predators ("mesopredators," such as raccoons, skunks, and domestic cats), modified stream morphology, reduced carbon sequestration, and increased incidence of disease (Ripple et al. 2014). Large herbivores also play keystone roles in many ecosystems (Fuhlendorf et al. 2009; Daskin et al. 2016; Lundgren et al. 2020; Karp et al. 2021; Ratajczak et al. 2022; Pedersen et al. 2023), as do some smaller engineering herbivores such as beaver (Castro et al. 2015) and gopher tortoise (*Gopherus polyphemus*) (Eisenberg 1983). For these various reasons, conservation planners typically use large mammals, as well as ecological engineers, as focal species for conservation planning. Restoration of ecologically effective, not just minimally viable, populations of such strongly interacting species to modern landscapes is a fundamental goal of rewilding (Soulé et al. 2003; Svenning et al. 2016; Ripple et al. 2022; Mungi et al. 2023).

LANDSCAPE ECOLOGY

Landscape ecology developed within the fields of geography and vegetation science in central Europe, with the first use of the term by the German geographer Carl Troll (1939), who wished to stimulate cooperation between geographers and biologists in the interpretation of aerial photography. From the beginning landscape ecology was a more empirical and descriptive science than metapopulation biology and less concerned with theory. Whereas metapopulation and island biogeographic models usually represent landscapes unrealistically as discrete patches of suitable habitat embedded in a homogeneous matrix of unsuitable habitat, landscape ecology has as its starting point "a detailed description of the often complex structure of real landscapes" (Hanski and Gaggiotti 2004). The promise of landscape ecology for metapopulation biology is to incorporate the complex mosaics of real landscapes into metapopulation models (Hanski and Simberloff 1997; With 2004).

Landscape ecology was imported to North America from Europe chiefly by Harvard ecologist Richard T. T. Forman (Forman 1981; Forman and Godron 1981, 1986; Forman 1995). The influential paper by Forman and Godron (1981) was descriptive, classifying landscape components such as patches, corridors, and matrix and describing their interactions. Soon, however, the field became more dynamic, considering movements of individuals and population processes across heterogeneous landscapes, complex disturbance regimes, and shifting mosaics of habitat patches over time and space (Risser et al. 1984; Forman 1987, 1995; Turner 1989; Wiens et al. 1993). The reciprocal interactions between spatial patterns and ecological processes and the fundamental role of spatial heterogeneity became central concerns of landscape ecology (Turner 2005), aided by advancements in spatial statistics and GIS. The relevance of landscape patterns and processes to biodiversity conservation was noted earlier, with Pickett and Thompson (1978) describing how nature reserves often need to be very large to incorporate disturbance-generated patch dynamics and maintain internal recolonization sources for disturbed patches.

Professor Larry Harris, a wildlife ecologist at the University of Florida, and his PhD student Reed Noss were both inspired by landscape ecology. Noss (1983b) introduced a "regional landscape approach to maintain diversity," which emphasized consideration of species diversity patterns across large landscapes rather than site by site (which was the typical approach) and urged creation of regional reserve networks with a mix of large and small reserves connected by habitat corridors. Harris (1984) employed several concepts from landscape ecology, along with options for countering fragmentation in managed forests (Harris 1984). Harris's strategy emphasized maintenance of the characteristic regional fauna, rather than simple species richness. His book, *The Fragmented Forest*, focused on the Oregon Cascades and recommended conservation of existing old-growth stands, an integrated system of long-rotation forest patches surrounded by replacement stands, and a dendritic network of habitat corridors with embedded small to medium-sized forest patches connecting large forest patches.

Harris (1984) also introduced the concept of the multiple-use module (MUM), a protected core area surrounded by buffer zones with intensity of protection increasing inward and intensity of human uses increasing outward. The MUM concept was further elaborated by Noss and Harris (1986), who noted that MUMs are consistent with agency mandates for multiple uses yet protect the most valuable and sensitive sites from human disturbance. A combination of MUMs and broad habitat corridors serves as a general spatial model for regional reserve networks (Noss 1992; fig. 2.1a). In a landscape that is still largely undeveloped, the model can be inverted with a "reverse matrix" of wildland and the core areas consisting of cities and other human settlements and the corridors as human transportation corridors (fig. 2.1b).

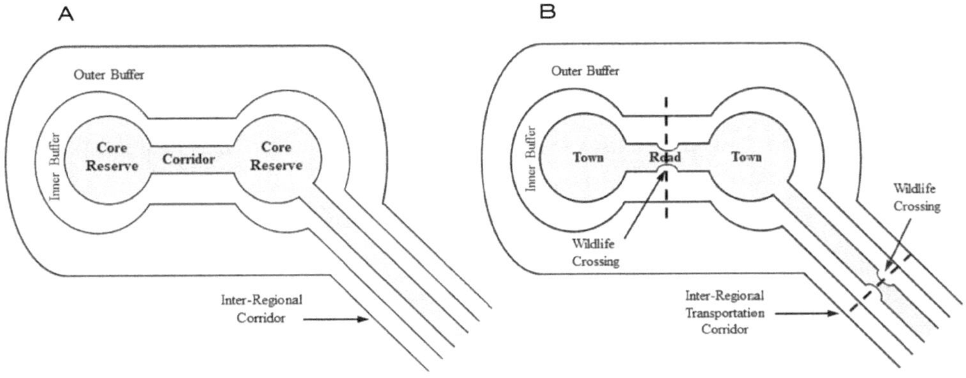

FIGURE 2.1. (*a*) A core-buffer-corridor model for human-dominated landscapes, with multiple-use modules (cores and buffers) connected by broad habitat corridors to create a regional reserve network (Noss 1992). (*b*) An inverted (reverse matrix) model for relatively undeveloped landscapes, with the cores consisting of human settlements buffered from the wildland and connected by human transportation corridors with wildlife crossings.

Early History of the Florida Wildlife Corridor

The Florida Wildlife Corridor is an example of ambitious landscape conservation planning on a statewide scale. The concept of a statewide ecological corridor has a long history in Florida. It is a planning model that originated largely in Florida and has inspired similar designs across North America and beyond (see chapter 10). As discussed earlier, the genesis of the Florida Wildlife Corridor in the early 1980s reflects several trends in the science of ecology and in the conservation movement that originated in the twentieth century and continue today.

While he was finishing writing *The Fragmented Forest* back home in Florida after his sabbatical in Oregon, Larry Harris began identifying specific "landscape linkages" to tie together large conservation areas, such as national forests and national wildlife refuges, within Florida and between Florida and adjacent states (Harris 1988; Harris and Gallagher 1989). These linkages included the Pinhook Swamp to connect Osceola National Forest to Okefenokee National Wildlife Refuge in northern peninsular Florida and adjacent Georgia; the Yellow River Ravines between Eglin Air Force Base and Blackwater River State Forest in the western Panhandle; connections between St. Marks National Wildlife Refuge and Apalachicola National Forest in the eastern Panhandle; the Wekiva River corridor between Wekiwa Springs State Park and Ocala National Forest in Central Florida; and connections among Loxahatchee National Wildlife Refuge, Jonathan Dickinson State Park, and other existing or potential conservation areas in southeastern Florida (Hoctor et al. 2015). All the landscape linkages recognized and advocated by Harris are now part of the modern Florida Wildlife Corridor, and most of them are partially to fully protected. Initially rejected as unrealistic, they were ultimately accepted and implemented.

Reed Noss, while working for the Ohio Department of Natural Resources (ODNR) in the early 1980s, independently developed a proposal for a regional network of core areas, corridors, and buffer zones connecting and surrounding federal, state, and private conservation areas in southern Ohio (Noss 1983a, 1987). Although rejected by the ODNR, portions of that proposed network were later implemented by TNC's Ohio chapter, especially the high-priority connections among its Edge of Appalachia preserves and to the nearby Shawnee State Forest. After moving to Florida in 1983, Noss collaborated with Harris to produce a regional network design for northern peninsular Florida, centered on the Suwannee River (from its origin in the Okefenokee National Wildlife Refuge) and its tributaries and extending to existing and proposed conservation

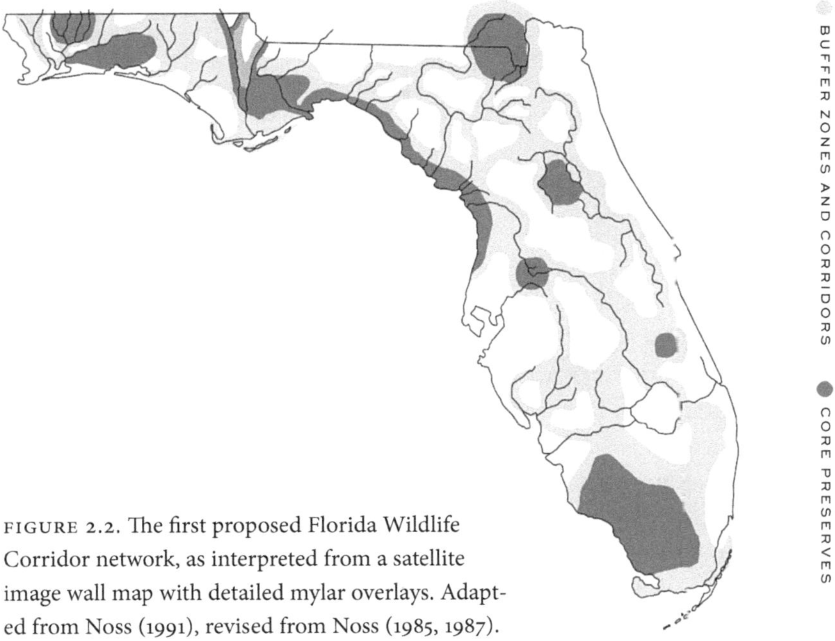

FIGURE 2.2. The first proposed Florida Wildlife Corridor network, as interpreted from a satellite image wall map with detailed mylar overlays. Adapted from Noss (1991), revised from Noss (1985, 1987).

areas along the Big Bend Coast of the Gulf of Mexico (Noss and Harris 1986). Concurrently, Noss developed the first statewide Florida Wildlife Corridor (see the section "Wilderness Recovery and Rewilding") (Noss 1985, 1987, 1991, 1992; Noss and Cooperrider 1994) (fig. 2.2). The early Florida Wildlife Corridor was introduced to the public in a Florida Wildlife Corridors conference organized by Bill Partington of the Florida Conservation Foundation in the spring of 1986. The conference was well attended and was covered by many newspapers, with the statewide corridor map appearing on some front pages and generating both positive and negative reactions from the public.

Beyond the attendees at the 1986 Corridors conference, the early Florida Wildlife Corridor (the Corridor) was not widely embraced because it was considered too radical and impractical. The mainstream environmental groups in Florida either opposed or expressed no interest in the Corridor at that time. Some ecologists opposed the corridor strategy outright, warning of potential negative effects such as the spread of diseases (Simberloff and Cox 1987; Simberloff et al. 1992). Agricultural industry groups, which enthusiastically support the Corridor today, vigorously opposed it in the 1980s. A likely explanation for this change of sentiment is that the early Corridor planning in Florida offered no explicit incentives for private landowners, whereas the modern Florida Wildlife Corridor does in the form of voluntary conservation easements for ranches and other private landholdings important for conservation (see chapter 9). Lacking a foundation of strong public support in the 1980s, the concept of a statewide corridor network for Florida was too far ahead of its time and was largely forgotten for nearly a decade.

Still, by the late 1980s Florida was well primed for ambitious conservation actions, having passed and implemented a series of significant regional planning initiatives beginning in the 1970s, including the creation of a statewide system of water management districts, landmark growth management legislation (later disabled by Governor Rick Scott), and several land acquisition programs, including Environmentally Endangered Lands, Save Our Rivers, Save Our Coasts, and the Conservation and Recreation Lands (CARL) program (Henderson 2022; and see chapter 3). Statewide conservation planning and land acquisition in Florida accelerated dramatically in the early 1990s after the passage of the Preservation 2000 (P-2000) Act in 1990 and the availability of significant funds ($3 billion over one decade) for conservation land acquisition (see chapter 3).

In 1991, the Conservation Fund and 1000 Friends of Florida initiated a Florida Greenways Program. A gubernatorial committee (the Florida Greenways Commission) was convened in 1993 and completed a report in 1994. Greenways legislation was adopted in 1997, the same year the first iteration of the Florida Ecological Greenways Network (FEGN) map was produced (Hoctor et al. 2000; see chapters 3 and 7). The FEGN represents a substantial refinement of the original Florida Wildlife Corridor (Noss 1985, 1987) and was based on Florida's growing wealth of high-quality spatial data on wildlife, water, and other natural resource conservation priorities and collaboration with a diverse group of experts. Nevertheless, to this day the basic pattern of the statewide corridor network has remained essentially the same since the mid-1980s. This is unsurprising because all planning efforts used Florida panther and Florida black bear as key focal species for landscape design and used river systems and other natural landscape features to connect existing or proposed conservation areas across the state. The amazingly successful history—so far—of the FEGN and the modern Florida Wildlife Corridor is the topic of chapter 7.

Literature Cited

Anonymous. 1980. Editorial. Earth First! 1(1):1.

Beier, P., and R. F. Noss. 1998. Do habitat corridors provide connectivity? Conservation Biology 12:1241–1252.

Bennett, A. F. 1998. Linkages in the Landscape: The Role of Corridors and Connectivity in Wildlife Conservation. IUCN, Gland, Switzerland, and Cambridge, UK.

Brown, J. H., and A. Kodric-Brown. 1977. Turnover rates in insular biogeography: Effect of immigration on extinction. Ecology 58:445–449.

Burgess, R. L., and D. M. Sharpe, editors. 1981. Forest Island Dynamics in Man-Dominated Landscapes. Springer-Verlag, New York.

Carver, S., I. Convery, S. Hawkins, R. Beyers, A. Eagle, Z. Kun, E. Van Maanen, Y. Cao, M. Fisher, S. Edwards, C. Nelson, G. Gann, S. Shurter, K. Aguilar, B. Ripple, J. Davis, T. Sinclair, M. Bekoff, R. Noss, D. Foreman, et al. 2021. Guiding principles for rewilding. Conservation Biology 35:1882–1893. https://doi.org/10.1111/cobi.13730.

Castro, J., M. Pollock, C. Jordan, G. Lewallen, and K. Woodruff. 2015. The Beaver Restoration Guidebook: Working with Beaver to Restore Streams, Wetlands, and Floodplains. Version 1.02. United States Fish and Wildlife Service. https://www.fws.gov/media/beaver-restoration-guidebook.

Cox, J., R. Kautz, M. MacLaughlin, and T. Gilbert. 1994. Closing the Gaps in Florida's Wildlife Habitat Conservation System. Florida Game and Fresh Water Fish Commission, Tallahassee.

Croker, R. A. 1991. Pioneer Ecologist: The Life and Work of Victor Ernest Shelford 1877–1968. Smithsonian Institution Press, Washington, DC.

Croker, R. A. 2001. Stephen Forbes and the Rise of American Ecology. Smithsonian Institution Press, Washington, DC.

Curtis, J. T. 1956. The modification of mid-latitude grasslands and forests by man. Pages 721–736 in W. L. Thomas, editor. Man's Role in Changing the Face of the Earth. University of Chicago Press, Chicago.

Damschen, E. I., N. M. Haddad, J. L. Orrock, J. J. Tewksbury, and D. J. Levey. 2006. Corridors increase plant species richness at large scales. Science 313:1284–1286.

Daskin, J. H., M. Stalmans, and R. M. Pringle. 2016. Ecological legacies of civil war: 35-year increase in savanna tree cover following wholesale large-mammal declines. Journal of Ecology 104:79–89.

Diamond, J. M. 1975. The island dilemma: Lessons of modern biogeographic studies for the design of natural preserves. Biological Conservation 7:129–146.

Dinerstein, E., D. Olson, A. Joshi, C. Vynne, N. D. Burgess, E. Wikramanayake, N. Hahn, S. Palminteri, P. Hedao, R. Noss, et al. 2017. An ecoregion-based approach to protecting half the terrestrial realm. BioScience 67:534–545.

Dinerstein, E., C. Vynne, E. Sala, A. Joshi, S. Fernando, T. E. Lovejoy, J. Mayorga, D.

M. Olson, G. P. Asner, J. A. Baillie, N. D. Burgess, K. Burkart, R. F. Noss, et al. 2019. A global deal for nature: Guiding principles, milestones, and targets. Science Advances 5: Eaaw2869.
Duncan, D. 2009. George Melendez Wright and the national park idea. George Wright Forum 26(1):4–13.
Eisenberg, J. 1983. The gopher tortoise as a keystone species. Pages 1–4 in R. J. Bryant and R. Franz, editors. The Gopher Tortoise: A Keystone Species. Proceedings of the 4th Annual Meeting of the Gopher Tortoise Council. Florida State Museum, Gainesville.
Estes, J. A., J. Terborgh, J. S. Brashares, et al. 2011. Trophic downgrading of Planet Earth. Science 333:301–306.
Fahrig, L. 2017. Ecological responses to habitat fragmentation per se. Annual Review of Ecology, Evolution, and Systematics 48:1–23.
Fletcher, R. J., Jr., R. K. Didham, C. Banks-Leite, J. Barlow, R. M. Ewers, J. Rosindell, R. D. Holt, A. Gonzalez, R. Pardini, E. I. Damschen, F. P. L. Melo, L. Ries, J. A. Prevedello, T. Tscharntke, W. F. Laurance, T. Lovejoy, and N. M. Haddad. 2018. Is habitat fragmentation good for biodiversity? Biological Conservation 226:9–15.
Foley, M. M., et al. 2017. Dam removal: Listening in. Water Resources Research 53:5229–5246, https://doi.org/10.1002/2017WR020457.
Foote, J. 1990. Trying to take back the planet. Newsweek 115(6):24–25.
Forbes, S. A. 1887 (reprinted 1925). The lake as a microcosm. Bulletin of the Illinois Natural History Survey 15:537–550.
Foreman, D. 1981. Earth First! The Progressive 45 (October):39–42.
Forman, R. T. T. 1981. Interactions among landscape elements: A core of landscape ecology. Pages 35–48 in Perspectives in Landscape Ecology. Proceedings of the International Congress of Landscape Ecology. Veldhoven Pudoc Publishers, Wageningen, The Netherlands.
Forman, R. T. T. 1987. The ethics of isolation, the spread of disturbance, and landscape ecology. Pages 213–229 in M. G. Turner, editor. Landscape Heterogeneity and Disturbance. Springer-Verlag, New York.
Forman, R. T. T. 1995. Land Mosaics: The Ecology of Landscapes and Regions. Cambridge University Press, Cambridge, UK.
Forman, R. T. T., and M. Godron. 1981. Patches and structural components for a landscape ecology. BioScience 31:733–740.
Forman, R. T. T., and M. Godron. 1986. Landscape Ecology. John Wiley and Sons, New York.
Fuhlendorf, S. D., D. M. Engle, J. Kerby, and R. Hamilton. 2009. Pyric herbivory: Rewilding landscapes through the recoupling of fire and grazing. Conservation Biology 23:588–598.

Gilbert-Norton, L., R. Wilson, J. R. Stevens, and K. H. Beard. 2010. A meta-analytic review of corridor effectiveness. Conservation Biology 24:660–668.

Goncalves-Souza, T., J. M. Chase, N. M. Haddad, et al. 2025. Species turnover does not rescue biodiversity in fragmented landscapes. Nature 2025:1–5. https://doi.org/10.1038/s41586-025-08688-7.

Haddad, N. M., D. R. Bowne, A. Cunningham, B. J. Danielson, D. J. Levey, S. Sargent, and T. Spira. 2003. Corridor use by diverse taxa. Ecology 84:609–615.

Haddad, N. M., L. A. Brudvig, J. Clobert, K. F. Davies, A. Gonzalez, R. D. Holt, T. E. Lovejoy, J. O. Sexton, M. P. Austin, C. D. Collins, W. M. Cook, E. I. Damschen, R. M. Ewers, B. L. Foster, C. N. Jenkins, A. J. King, W. F. Laurance, D. J. Levey, C. R. Margules, B. A. Melbourne, A. O. Nicholls, J. L. Orrock, D.-X. Song, and J. R. Townsend. 2015. Habitat fragmentation and its lasting impact on Earth's ecosystems. Science Advances 1: e1500052.

Hanski, I. 1998. Metapopulation dynamics. Nature 396:41–49.

Hanski, I. 1999. Metapopulation Ecology. Oxford University Press, Oxford, UK.

Hanski, I., and O. E. Gaggiotti. 2004. Metapopulation biology: Past, present, and future. Pages 3–22 in I. Hanski and O. E. Gaggiotti, editors. Ecology, Genetics, and Evolution of Metapopulations. Elsevier, Amsterdam, the Netherlands.

Hanski, I., and D. Simberloff. 1997. The metapopulation approach, its history, conceptual domain, and application to conservation. Pages 5–26 in I. A. Hanski and M. E. Gilpin, editors. Metapopulation Biology: Ecology, Genetics, and Evolution. Academic Press, London.

Harris, L. D. 1984. The Fragmented Forest. Island Biogeographic Theory and the Preservation of Biotic Diversity. University of Chicago Press, Chicago.

Harris, L. D. 1988. Landscape linkages: The dispersal corridor approach to wildlife conservation. Transactions of the North American Wildlife and Natural Resources Conference 53:595–607.

Harris, L. D., and P. B. Gallagher. 1989. New initiatives for wildlife conservation. The need for movement corridors. Pages 11–34 in G. Mackintosh, editor. In Defense of Wildlife: Preserving Communities and Corridors. Defenders of Wildlife, Washington, DC.

Harris, L. D., and J. Scheck. 1991. From implications to applications: The dispersal corridor principle applied to the conservation of biological diversity. Pages 189–220 in D. A. Saunders and R. J. Hobbs, editors. Nature Conservation 2: The Role of Corridors. Surrey Beatty & Sons, Chipping Norton, New South Wales.

Harrison, S. 1991. Local extinction in a metapopulation context: An empirical evaluation. Biological Journal of the Linnean Society 42:73–88.

Henderson, C. 2022. Forces of Nature: A History of Florida Land Conservation. University Press of Florida, Gainesville.

Hoctor, T. S., W. L. Allen III, M. H. Carr, P. D. Zwick, E. Huntley, D. J. Smith, D. S. Maehr, R. Buch, and R. Hilsenbeck. 2008. Land corridors in the Southeast USA: Connectivity to protect biodiversity and ecosystem services. Journal of Conservation Planning 4:90–122.

Hoctor, T. S., M. H. Carr, and P. D. Zwick. 2000. Identifying a linked reserve system using a regional landscape approach: The Florida Ecological Network. Conservation Biology 14:984–1000.

Hoctor, T., R. Noss, R. Hilsenbeck, J. Guthrie, and C. Ward. 2015. The history of Florida Wildlife Corridor science and planning efforts. https://floridawildlifecorridor.org/wp-content/uploads/2011/12/FWC_History_11_09_2015.pdf. Accessed August 24, 2024.

Jenkins, R. E. 1978. Heritage classification: The elements of ecological diversity. Nature Conservancy News 28(1):24–25, 30.

Jenkins, R. E. 1985. Information methods: Why the heritage programs work. Nature Conservancy News 35(6):21–23.

Jepson, P. 2025. De-extinction beyond species: Restoring ecosystem functionality through large herbivore rewilding. Cambridge Prisms: Extinction 2025:1–14. https://doi.org/10.1017/ext.2024.27.

Jongman, R. H. G., and G. Pungetti. 2004. Ecological Networks and Greenways: Concept, Design, Implementation. Cambridge University Press, Cambridge, UK.

Karp, A. T., J. T. Faith, J. R. Marlon, and A. C. Staver. 2021. Global response of fire activity to late Quaternary grazer extinctions. Science 374:1145–1148.

Kendeigh, S. C., H. I. Baldwin, V. H. Cahalane, C. H. D. Clarke, C. Cottam, W. P. Cottam, I. McT. Cowan, P. Dansereau, J. H. Davis, F. W. Emerson, L. T. Haig, A. Hayden, C. L. Hayward, J. M. Linsdale, J. A. MacNab, and J. E. Potzger. 1950–1951. Nature sanctuaries in the United States and Canada: A preliminary inventory. The Living Wilderness 15(35):1–45.

Levins, R. 1970. Extinction. Lectures and Notes in Mathematics 2:75–107.

Locke, H. 2013. Why nature needs half: A necessary and hopeful agenda for parks and protected areas. Parks: The International Journal for Parks and Protected Areas 19(2):9–18.

Locke, H. 2014. Nature needs half: A necessary and hopeful agenda for protected areas in North America and around the world. The George Wright Forum 31:359–371.

Locke, H. 2018. The International Movement to Protect Half the World: Origins, Scientific Foundations, and Policy Implications. Earth Systems and Environmental Sciences Reference Manual. Elsevier, Amsterdam, The Netherlands.

Lundgren, E. J., D. Ramp, J. Rowan, et al. 2020. Introduced herbivores restore Late Pleistocene ecological functions. Proceedings of the National Academy of Sciences 117:7871–7878.

MacArthur, R. H., and E. O. Wilson. 1967. The Theory of Island Biogeography. Princeton University Press, Princeton, New Jersey.

Mungi, N. A., Y. V. Jhala, Q. Quereshi, E. le Roux, and J.-C. Svenning. 2023. Megaherbivores provide biotic resistance against alien plant dominance. Nature Ecology and Evolution. https://doi.org/10.1038/s41559-023-02181-y.

Noss, R. F. 1983a. Re-creating an Ohio Valley wilderness. Midwest Earth Advocate 1(1):4.

Noss, R. F. 1983b. A regional landscape approach to maintain diversity. BioScience 33:700–706.

Noss, R. F. 1985. Wilderness recovery and ecological restoration: An example for Florida. Earth First! 5(8)18–19.

Noss, R. F. 1986. Recipe for wilderness recovery. Earth First! 6(8):22, 25.

Noss, R. F. 1991. Wilderness recovery: Thinking big in restoration ecology. The Environmental Professional 13:225–234.

Noss, R. F. 1992. The Wildlands Project: Land conservation strategy. Wild Earth (Special Issue): 10–25.

Noss, R. F. 1996. Protected areas: How much is enough? Pages 91–120 in R. G. Wright, editor. National Parks and Protected Areas: Their Role in Environmental Protection. Blackwell, Cambridge, Massachusetts.

Noss, R. F. 2020. The spectrum of wildness and rewilding: Justice for all. Pages 167–182 in H. Kopnina and H. Washington, editors. Conservation—Integrating Social and Ecological Justice. Springer, The Hague, the Netherlands.

Noss, R. F., and A. Cooperrider. 1994. Saving Nature's Legacy: Protecting and Restoring Biodiversity. Island Press, Washington, DC.

Noss, R. F., B. Csuti, and M. J. Groom. 2006. Habitat fragmentation. Pages 213–251 in M. J. Groom, G. K. Meffe, and R. C. Carroll, editors. Principles of Conservation Biology, Third edition. Sinauer Associates, Sunderland, Massachusetts.

Noss, R. F., A. P. Dobson, R. Baldwin, P. Beier, C. R. Davis, D. A. DellaSala, J. Francis, H. Locke, K. Nowak, R. R. Lopez, C. Reining, S. Trombulak, and G. Tabor. 2012. Bolder thinking for conservation. Conservation Biology 26:1–4.

Noss, R. F., and L. D. Harris. 1986. Nodes, networks, and MUM's: Preserving diversity at all scales. Environmental Management 10:299–309.

Noss, R. F., M. A. O'Connell, and D. D. Murphy. 1997. The Science of Conservation Planning: Habitat Conservation under the Endangered Species Act. Island Press, Washington, DC.

Odum, E. P., and H. T. Odum. 1972. Natural areas as necessary components of man's total environment. Transactions of the North American Wildlife and Natural Resources Conference 37:178–189.

Pearson, A. M. 2017. Force of Nature: George Fell, Founder of the Natural Areas Movement. University of Wisconsin Press, Madison.

Pedersen, R. O., S. Faurby, and J.-C. Svenning. 2023. Late-Quaternary megafauna extinctions have strongly reduced mammalian vegetation consumption. Global Ecology and Biogeography 2023:00:1–13.

Pickett, S. T. A., and J. N. Thompson. 1978. Patch dynamics and the size of nature reserves. Biological Conservation 13:27–37.

Pressey, R. L., R. M. Cowling, and M. Rouget. 2003. Formulating conservation targets for biodiversity pattern and process in the Cape Floristic Region, South Africa. Biological Conservation 112:99–127.

Prugh, L. A., K. E. Hodges, A. R. E. Sinclair, and J. S. Brashares. 2008. Effect of habitat area and isolation on fragmented animal populations. Proceedings of the National Academy of Sciences 105:20770–20775.

Ratajczak, Z., S. L. Collins, J. M. Blair, et al. 2022. Reintroducing bison results in long-running and resilient increases in grassland diversity. Proceedings of the National Academy of Sciences 119(36):e2210433119.

Ripple, W. J., J. A. Estes, R. L. Beschta, et al. 2014. Status and ecological effects of the world's largest carnivores. Science 343:1241484-1–1241484-11.

Ripple, W. J., M. K. Phillips, C. Wolf, R. L. Beschta, J. Vucetich, J. B. Kauffman, B. E. Law, A. Wirsing, J. E. Lambert, E. Leslie, C. Vynne, E. Dinerstein, R. Noss, G. Wuerthner, D. A. DellaSala, J. T. Bruskotter, M. P. Nelson, E. Crist, and D. M. Ashe. 2022. Rewilding the American West. BioScience. https://doi.org/10.1093/biosci/biac069.

Risser, P. G., J. R. Karr, and R. T. T. Forman. 1984. Landscape ecology: Directions and approaches. Special Publication No. 2, Illinois Natural History Survey, Champaign.

Sayen, J. 1987. The Appalachian Mountains vision and wilderness. Earth First! 7(5):26–30.

Sellars, R. W. 2000. The significance of George Wright. George Wright Forum 17(4):46–50.

Shelford, V. E., editor. 1926. Naturalist's Guide to the Americas. Committee on the Preservation of Natural Conditions of the Ecological Society of America. Williams and Wilkins, Baltimore, Maryland.

Shelford, V. E. 1933a. The preservation of natural biotic communities. Ecology 14:240–245.

Shelford, V. E. 1933b. Conservation versus preservation. Science 77:535.

Shelford, V. E. 1944. The conflict between science and biological industry. Science 100:450–451.

Simberloff, D., and J. Cox. 1987. Consequences and costs of conservation corridors. Conservation Biology 1:63–71.

Simberloff, D., J. A. Farr, J. Cox, and D. W. Mehlman. 1992. Movement corridors: Conservation bargains or poor investments? Conservation Biology 6:493–504.

Soulé, M. E., J. A. Estes, J. Berger, and C. Martinez del Rio. 2003. Recovery goals for ecologically effective numbers of endangered keystone species. Conservation Biology 17:1238–1250.

Soulé, M., and R. Noss. 1998. Rewilding and biodiversity: Complementary goals for continental conservation. Wild Earth 8(3):18–28.

Sutherland, C., D. A. Elston, and X. Lambin. 2012. Multi-scale processes in metapopulations: Contributions of stage structure, rescue effect, and correlated extinctions. Ecology 93:2465–2473.

Svenning, J.-C., P. B. M. Pedersen, C. J. Donlan, et al. 2016. Science for a wilder Anthropocene: Synthesis and future directions for trophic rewilding research. Proceedings of the National Academy of Sciences 113:898–906.

Tansley, A. G. 1935. The use and abuse of vegetational concepts and terms. Ecology 16:284–307.

Troll, C. 1939. Luftbildplan und ökologische Bodenforschung. Zeitschrift der Gesellschaft für Erdkunde zu Berlin 1939:241–298.

Trombulak, S. C., and R. F. Baldwin, editors. 2010. Landscape-Scale Conservation Planning. Springer, New York.

Turner, M. G. 1989. Landscape ecology: The effect of pattern on process. Annual Review of Ecology and Systematics 20:171–197.

Turner, M. G. 2005. Landscape ecology: What is the state of the science? Annual Review of Ecology, Evolution, and Systematics 36:319–344.

United Nations Environment Programme (UNEP). 2022. UN Biodiversity Conference (COP 15). https://www.unep.org/un-biodiversity-conference-cop-15.

Wegner, J. F., and G. Merriam. 1979. Movements by birds and small mammals between a wood and adjoining farmland habitat. Journal of Applied Ecology 16:349–357.

Wiens, J. A., N. C. Stenseth, B. Van Horne, and R. A. Ims. 1993. Ecological mechanisms and landscape ecology. Oikos 66:369–380.

Wilson, E. O. 2016. Half-Earth: Our Planet's Fight for Life. Liveright, New York.

Wilson, E. O., and E. O. Willis. 1975. Applied biogeography. Pages 522–534 in M. L. Cody and J. M. Diamond, editors. Ecology and Evolution of Communities. Belknap Press of Harvard University Press, Cambridge, Massachusetts.

With, K. A. 2004. Metapopulation dynamics: Perspectives from landscape ecology. Pages 23–44 in I. Hanski and O. E. Gaggiotti, editors. Ecology, Genetics, and Evolution of Metapopulations. Elsevier, Amsterdam, the Netherlands.

Wolke, H. 1983. Editorial: RARE II. Earth First! 3(3):3.

Wooding, J. B., and T. S. Hardisky. 1994. Home range, habitat use, and mortality of black bears in north-central Florida. Pages 349–356 in Bears: Their Biology and Management, Part 1: A Selection of Papers from the Ninth International Conference on Bear Research and Management, Missoula, Montana, February 23–28, 1992.

3

Prioritization Methods for Ambitious Land Conservation under Preservation 2000, Florida Forever, and the Rural and Family Lands Protection Program

Jon Oetting, Amy Knight, and Thomas S. Hoctor

Introduction

A researcher wishing to examine the effectiveness of conservation tools available to the profession, from applied conservation research to strategic planning, from fee simple and less-than-fee land acquisition to landowner incentives, technical support programs, and land management techniques, could hardly find a more rewarding laboratory than the state of Florida over the past 50 years and beyond. By both necessity and opportunity, Florida has been at the forefront of both conservation theory and practice. Fortunate to remain largely undeveloped relative to the eastern United States for its first century or more of statehood, Florida saw exponential growth and development through the mid- to late twentieth century. Faced with rapid and tangible loss of natural places, Floridians coalesced around conserving the natural environment and have enthusiastically supported land acquisition for this purpose through multiple governors and legislatures controlled by both parties. Conservationists, scientists, and planners have responded to the need for coordinating land acquisition efforts with some of the most innovative conservation planning tools produced. This chapter tells the story of those efforts and tools.

Timeline of Land Acquisition Programs in Florida

Florida has a long history of acquiring land for the conservation of natural resources (Henderson 2022). The country's first national wildlife refuge, Pelican Island, was acquired in 1903 north of Vero Beach in Indian River County. Early acquisitions were primarily by the federal government, culminating in Everglades National Park in 1947. Prior to the 1960s, state efforts were largely opportunistic, including notable sites like Torreya State Park in 1935, Babcock-Webb Wildlife Management Area in 1941, and Withlacoochee State Forest in 1958 (Knight et al. 2011). In 1963 Florida's legislature enacted the Outdoor Recreation and Conservation Program (ORCP), which focused on state parks and recreation areas and was guided by a statewide outdoor recreation plan (SCORP), a planning process that has continued to the present (FDEP 2024a). Approximately 25,500 ha (68,000 ac) were acquired from 1963 to 1985 through this initial program. The Environmentally Endangered Lands (EEL) program was established in 1972 to focus on ecologically unique and threatened lands. EEL represented a significant ramp-up of funding and effort, resulting in nearly 152,000 ha (375,000 ac) acquired between 1972 and 1978. EEL was replaced by the Conservation and Recreation Lands (CARL) program in 1979, with around 104,000 ha (256,000 ac) conserved through 1990. Save Our Rivers and Save Our Coasts, each with a more specific conservation focus, were both added in 1981, conserving nearly 213,000 ha (527,000 ac) and 32,000 ha (78,000 ac), respectively, through 1990.

The year 1990 saw a landmark conservation achievement on a national scale when the Florida legislature enacted Preservation 2000 (P-2000), a 10-year, $3 billion land conservation program. P-2000 distributed 50% of funds through the CARL program, 30% through the state's five water management districts (a continuation of Save Our Rivers), 10% to Florida Communities Trust, a new program focusing on local parks and recreation areas, 1.3% to recreational trails, and the remainder to purchase inholdings and make small additions to existing conservation lands (Farr and Brock 2006). By all accounts P-2000 was a tremendous success, acquiring more than 720,000 ha (1.78 million ac) between 1990 and 2001. By the end of the decade, there was sufficient momentum for a successor program, and Florida Forever was enacted in 2001 for an additional

10 years and an expected $3 billion. The Florida Forever Act provided the most detailed guidance from the legislature to date as to the specific natural resource values and criteria that should be considered in prioritizing projects for acquisition, including an increased focus on the purchase of less-than-fee conservation easements on private lands. The appointed council overseeing acquisitions was expanded to include governor appointees along with agency heads and renamed the Acquisition and Restoration Council (ARC). Each December, the ARC ranks all Florida Forever projects among six funding categories: Critical Natural Lands, Partnerships and Regional Incentives, Less-Than-Fee, Climate Change Lands, Critical Historical Resources, and Substantially Complete. Actual funding has fluctuated over the years in response to economic and political conditions, with less than $100 million appropriated each year from 2010 through 2021 (FDEP 2024b). The Florida Forever program was extended indefinitely in 2010 and continues to the present. To date more than 375,000 ha (927,000 ac) have been purchased through Florida Forever (FNAI 2023a).

Early Acquisition Planning Efforts in Florida

The first acquisition programs, including ORCP and EEL, relied on internal evaluations at the Florida Department of Natural Resources (now merged into the Department of Environmental Protection) and were based primarily on staff expert judgment (Farr and Brock 2006). A more structured evaluation process was established for CARL in 1979. Staff evaluated projects and made recommendations to an appointed advisory council made up of agency heads, who selected and ranked projects. Land acquisition "projects" are typically single sites that are a focus of acquisition efforts and may have single or multiple property owners. Projects are typically submitted via an application process and evaluated as a single unit. At the beginning, project evaluation was still largely qualitative, focused on identifying what staff considered "important natural areas" (G. Knight, pers. comm.). In 1981 the Florida Natural Areas Inventory (FNAI) was founded as the latest state natural heritage program in The Nature Conservancy's growing Natural Heritage Network across the US and later Canada. The mission of the Natural Heritage Network was to provide a scientific foundation for conservation decisions by gathering information on the locations of rare species and other conservation features (Jenkins 1978, 1985). From the beginning, FNAI's core contract was with the CARL program to support land acquisition efforts. FNAI's nascent database of rare species and natural community occurrences throughout

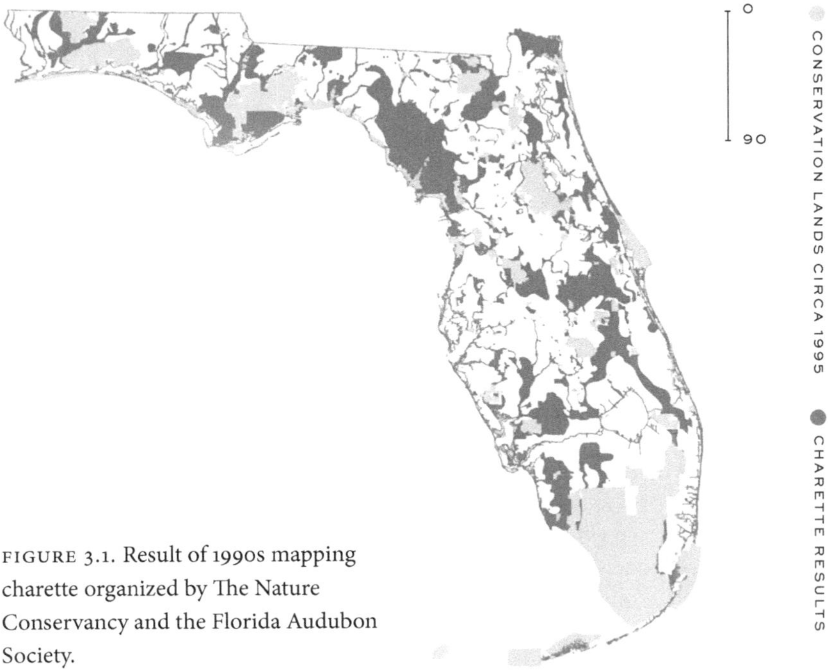

FIGURE 3.1. Result of 1990s mapping charette organized by The Nature Conservancy and the Florida Audubon Society.

the state became some of the earliest biological data available to acquisition staff to enable objective scientific evaluation of projects.

With the onset of P-2000, conservation planning began to play a more important role in the land acquisition process. In 1991 the Florida chapter of The Nature Conservancy, in partnership with the Florida Audubon Society and the Department of Natural Resources, convened a mapping charette to identify important conservation areas throughout the state (Hoctor et al. 2015). Conservationists and scientists developed a consensus map that focused on large natural and seminatural landscapes and riverine corridors (fig. 3.1). While landscape connectivity was not an explicit focus of the effort, the charette map did share similarities with the earlier statewide core areas and corridors map created by Reed Noss (see chapter 2).

In the mid-1990s, FNAI undertook a similar mapping effort to identify what they termed Areas of Conservation Interest (ACIs). FNAI scientists reviewed Florida Department of Transportation 1:25,000-scale aerial photos of the entire state and mapped assemblages of natural communities (fig. 3.2). The highest

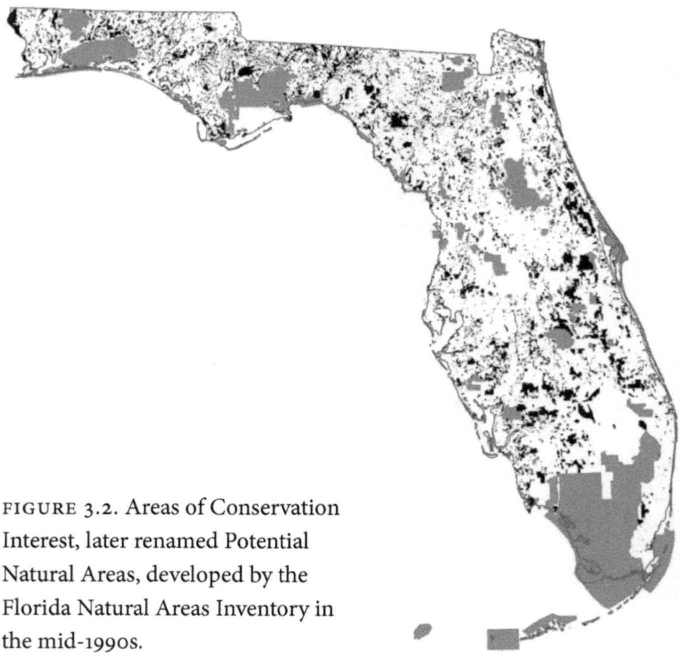

FIGURE 3.2. Areas of Conservation Interest, later renamed Potential Natural Areas, developed by the Florida Natural Areas Inventory in the mid-1990s.

priority ACIs included field-documented occurrences of rare species or natural communities, while lower priorities had specific acreage criteria for various community types. These maps were reviewed by local experts at a series of regional workshops around the state. Areas of Conservation Interest underwent several rounds of revision and eventually were renamed Potential Natural Areas (PNAs), a dataset that is still maintained by FNAI and updated periodically with current land cover.

The mid-1990s also saw the first GIS-based analysis of conservation needs in Florida. In 1994, Jim Cox, Randy Kautz, and colleagues at the Florida Game and Fresh Water Fish Commission (since renamed Florida Fish and Wildlife Conservation Commission) published *Closing the Gaps in Florida's Wildlife Habitat Conservation System*, a landmark study of habitat needs for 44 focal vertebrate species (Cox et al. 1994). They developed land cover–based habitat models for each species and applied basic population viability analysis to each model to examine which species had adequate protected habitat on existing conservation lands to sustain populations long-term. They estimated that 30 of the 44 species

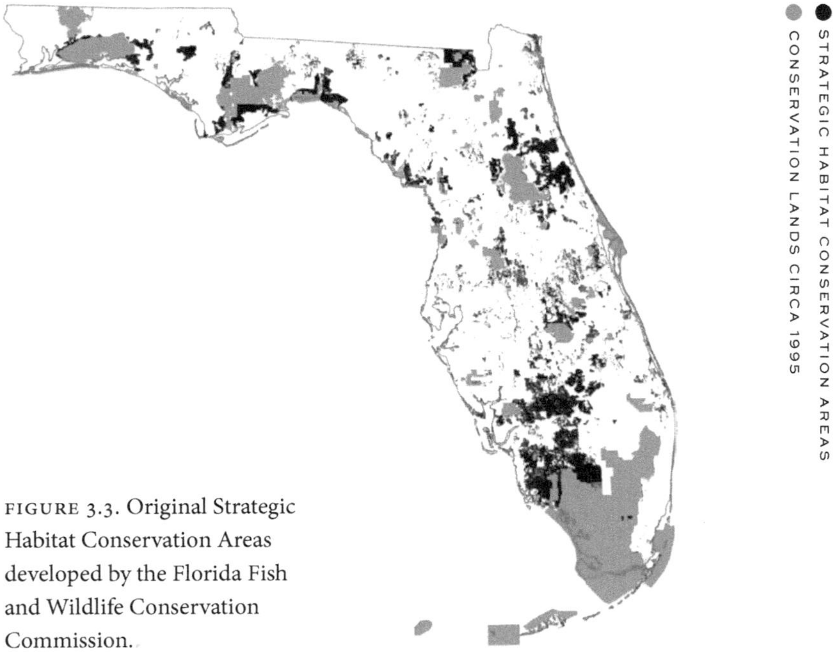

FIGURE 3.3. Original Strategic Habitat Conservation Areas developed by the Florida Fish and Wildlife Conservation Commission.

lacked adequate protection on conservation lands and identified for each species a series of Strategic Habitat Conservation Areas (SHCAs), additional habitat that needed protection to meet the long-term viability needs of each species (fig. 3.3). *Closing the Gaps* and its follow-up reports in the 2000s served as a national example for how state wildlife agencies could assess conservation needs at a statewide scale. All these maps—the charette, ACI/PNAs, and SHCAs—were referenced by land acquisition staff and helped guide P-2000 acquisition efforts.

At around the same time as the *Closing the Gaps* study, a new planning effort was beginning that would address landscape connectivity directly. In 1993, a governor-appointed Florida Greenways Commission was established to envision a statewide network of both recreational and ecological greenways (Hoctor et al. 2015). The Florida Greenways Commission was, in part, a culmination of work by Larry Harris and Reed Noss at the University of Florida in the 1980s to highlight the threat of habitat fragmentation and need for wildlife corridors (described in chapter 2) as well as a greenways program started and advocated by 1000 Friends of Florida and the Conservation Fund in 1990. The Florida Greenways Commission completed a report in 1994 (FGC 1994) that recommended establishment of a statewide greenways program. The Office of Greenways and Trails (OGT) was formed within the Florida Department of Environmental Protection (FDEP) to support the development of this new program. OGT contracted with a research team at the University of Florida in 1995 (including Peggy Carr, Paul Zwick, Tom Hoctor, and Mark Benedict) to develop the Florida Ecological Greenways Network; a separate recreational trails planning effort was also undertaken.

The University of Florida team conducted a GIS landscape connectivity analysis to identify key hubs and connecting corridors, with particular focus on connectivity for Florida black bear (*Ursus americanus floridanus*) and Florida panther (*Puma concolor*), large biodiversity hotspots, and surface water protection in intact watersheds. An important aspect of the Florida Ecological Greenways Network was the use of an expert technical advisory group (TAG), which was consulted extensively regarding both data sources and prioritization criteria used in the modeling. The first iteration of the Florida Ecological Greenways Network map was completed in 1997 and became part of the Florida Greenways Plan passed into state law that same year (Hoctor et al. 2000; Hoctor et al. 2008; see chapter 7 for a full account of the FEGN).

Florida Forever and Conservation Planning

By any measure, CARL and P-2000 had been conservation success stories. More than 809,000 ha (2 million ac) of natural Florida had been conserved. A broad range of resources was protected: some of Florida's last remaining undeveloped coastline, key concentrations of biodiversity, and river and spring buffers throughout the state are highlights. However, these programs lacked a clear, objective accounting of accomplishments. As momentum built around the efforts of The Nature Conservancy, Florida Audubon, and other NGOs for a successor program, Florida politics—like that of the nation—was placing increasing emphasis on government accountability. When the Florida Forever Act was passed in 2000, extensive and detailed guidelines and criteria were specified to ensure that the program was focusing on the right places and spending the public's money efficiently. Specifically, the act listed nine goals and 38 performance measures, along with several additional criteria to be considered in evaluating projects.

In addition to goals, performance measures, and criteria, the act spelled out overall guidelines for the direction of the program (Chapter 259.105(2) *F.S.*). Notable are the following:

FLORIDA FOREVER GOALS AND MEASURES

A. Enhance coordination and completion of projects

B. Increase the protection of Florida's biodiversity at the species, natural community, and landscape levels
  - acres acquired of significant strategic habitat conservation areas
  - acres acquired of highest priority conservation areas for Florida's rarest species
  - acres acquired of significant landscapes, landscape linkages, and conservation corridors, giving priority to completing linkages
  - acres acquired of underrepresented native ecosystems
  - landscape-sized protection areas of at least 50,000 acres that exhibit a mosaic of predominantly intact or restorable natural communities
  - increase in the number of occurrences of imperiled species

C. Protect, restore, and maintain the quality and natural functions of land, water, and wetland systems of the state
  - publicly owned land identified as needing restoration, enhancement, and management
  - acres acquired that protect natural floodplain functions
  - acres acquired that protect surface waters of the state
  - acres identified to minimize damage from flooding and the percentage acquired
  - acres acquired that protect fragile coastal resources
  - acres of functional wetland systems protected

D. Ensure that sufficient quantities of water are available to meet the current and future needs of natural systems and the citizens of the state
  - acres acquired of groundwater recharge areas critical to springs, sinks, aquifers, other natural systems, or water supply

E. Increase natural resource-based public recreational and educational opportunities

F. Preserve significant archaeological or historical sites

G. Increase the amount of forestland available for sustainable management of natural resources

H. Increase the amount of open space available in urban areas

I. Mitigate the effects of natural disasters and floods in developed areas
  - acres acquired within a 100-year floodplain or a coastal high hazard area

Additional Criteria

—project meets multiple goals
—enhances or facilitates management of properties already under public ownership
—project contributes to the solution of water resource problems on a regional basis
—project has a significant portion of its land area in imminent danger of development
—project may be acquired, in whole or in part, using alternatives to fee simple
—project is a joint acquisition, either among public agencies, nonprofit organizations, or private entities, or by a public-private partnership
—Projects for which the state's land conservation plans overlap with the military's need to protect lands, water, and habitat to ensure the sustainability of military missions
—Projects located within the Florida wildlife corridor*

Source: Chapter 259.105 *F.S.*
Not all measures and criteria listed in the act are included here.
*Added in a 2021 amendment to the act.

Acquisition of lands . . . shall be based on a comprehensive science-based assessment of the state's natural resources.

That assessment will "target essential conservation lands by prioritizing all current and future acquisitions based on a uniform set of data and planned so as to protect the integrity and function of ecological systems and working landscapes and provide multiple benefits."

While previous and existing state environmental programs have achieved varying degrees of success, few of these programs can be evaluated as to the extent of their achievements, primarily because performance measures, standards, outcomes, and goals were not established at the outset. Therefore, the Florida Forever program shall be developed and implemented in the context of measurable state goals and objectives.

> The Legislature recognizes that acquisition of lands in fee simple is only one way to achieve the aforementioned goals and encourages the use of less-than-fee interests, other techniques, and the development of creative partnerships between governmental agencies and private landowners.

These requirements would usher in a new era of conservation planning for Florida.

FLORIDA FOREVER CONSERVATION NEEDS ASSESSMENT AND F-TRAC

To address the requirements of the Florida Forever Act, the Florida Department of Environmental Protection, Division of State Lands, contracted FNAI in 2000 to develop the Florida Forever Conservation Needs Assessment (FFCNA). Under the leadership of Gary Knight, director, FNAI developed a series of statewide GIS maps addressing 16 of the measures listed in the Florida Forever Act. These measures were selected because of their focus on acquisition priorities (other measures focused on postacquisition activities such as restoration or management). Each map was developed based on best available statewide data and prioritized to inform acquisition needs; for a given resource, the mapping and prioritization methods were designed to be consistent across the state to ensure comparability among locations. FNAI established an Expert Working Group consisting of natural resource, conservation planning, and GIS modeling experts from state agencies, academia, and NGOs to review and advise all FFCNA work. Together they developed operational definitions for the resources addressed in each performance measure.

Some performance measures were straightforward—Strategic Habitat Conservation Areas are both a performance measure and an existing map—while others required considerable discussion and interpretation. For example, Significant Surface Waters were defined to address surface waters in relatively natural condition (not exceeding water quality thresholds or designated for restoration) and include Outstanding Florida Waters, shellfish harvesting areas, seagrass beds, Aquatic Preserves, National Wild and Scenic Rivers, springs, National Estuaries, Class 1 potable water supply sources, and water bodies deemed important for imperiled fish species (FNAI 2023b). FNAI modeled upland buffers for each of these groups of water bodies and prioritized them based on the potential an upland land acquisition had for influencing water quality (fig. 3.4). FNAI also expanded on the statewide database of rare species occurrences by producing

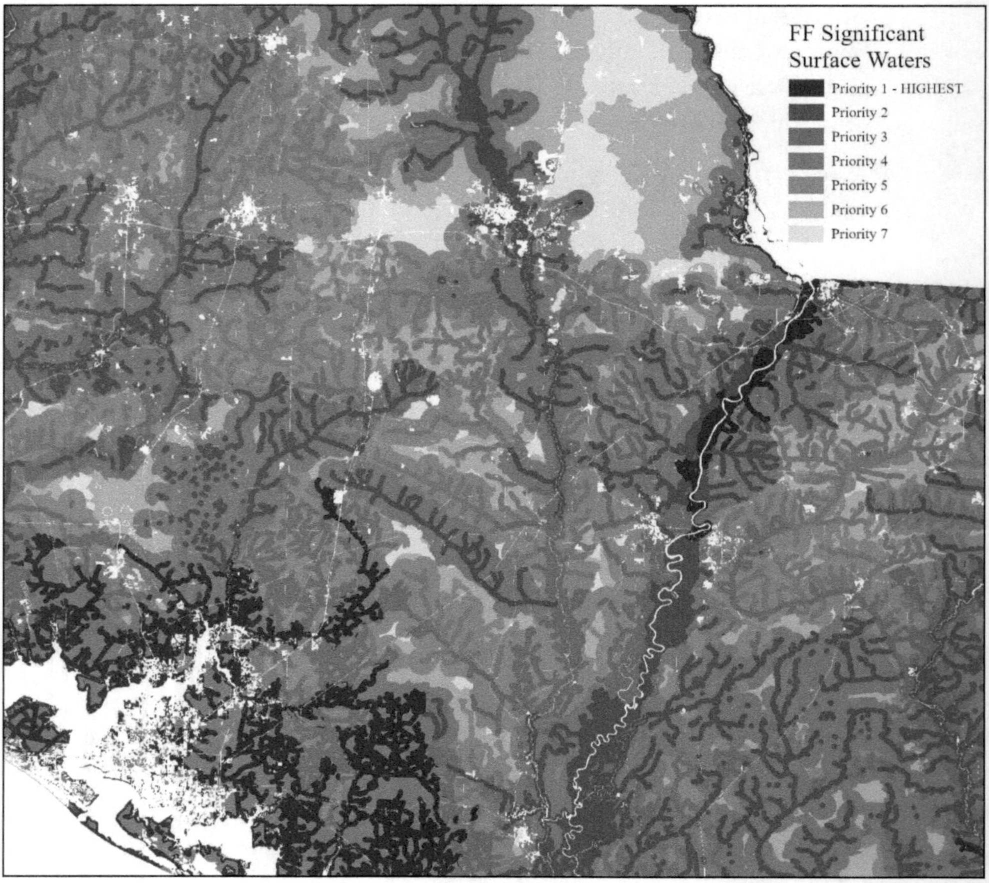

FIGURE 3.4. Detail of Significant Surface Waters priorities map from Florida Forever Conservation Needs Assessment. The Chipola River, located in the Florida Panhandle, is running north to south through the center of this map.

suitable habitat maps for more than 250 species (currently expanded to 400+), combined into the Rare Species Habitat Conservation Priorities map layer (all FFCNA maps can be viewed at https://geodata.fnai.org/).

The FFCNA maps serve multiple purposes. First, they identify a baseline for the amounts and locations of resources protected on conservation lands at the beginning of the Florida Forever program as well as those areas of high priority remaining to be acquired. With these baselines established, FNAI has tracked and reported progress of the Florida Forever program in acquiring additional resources (FNAI 2023a). The maps also form the basis of single-resource evaluations of Florida Forever projects and new proposals (FNAI 2023b), in which projects receive a score indicating how well they meet a given resource measure. To use the FFCNA data layers for analysis, the team, in collaboration with the Expert Working Group, first aggregates data layers that focus on similar resources. For single-resource evaluation, each project boundary is overlaid on each resource map, and acres of each resource priority class within the project are tabulated. A weighted average is then calculated and classed into a final score ranging from "very high" to "low/none." For some resource measures FNAI also identified minimum acreage threshold requirements for certain classes (for example, some amount of a resource's Priority 1 class might be required on-site for a project to score "very high"). The single-resource evaluation has the advantage of putting all projects on a level playing field, allowing decision-makers to compare projects of varying sizes and different locations around the state against a common dataset.

After developing the single-resource evaluation, the FFCNA Expert Working Group sought to provide a summary score for Florida Forever projects across most or all resource categories. The group decided on Marxan, a decision-support tool that uses an iterative site-selection algorithm to build a near-optimal portfolio of sites for acquisition (Ball et al. 2009). Marxan starts with a set of natural resource features distributed across the study area. Targets are set for the amount of each feature to conserve, and weights are assigned to prioritize some features over others. In this case the team used the FFCNA single-resource map layers as the resource features, with each priority class treated as a separate feature. The original analysis included the following resource layers: species, natural communities, surface waters, wetlands/floodplain, recharge, and forestry (but not ecological greenways; see the section "Landscape Connectivity"). The team also included a cost threshold (size limit) for the portfolio. Originally the cost threshold was based on an estimate of total acreage projected to be acquired by the Florida Forever program, but since the program has been extended indefinitely, the threshold is now set at 50% of the total acreage on the current list of Florida Forever projects. This analysis is termed F-TRAC, which loosely stands for Florida Forever Tool for Efficient Resource Acquisition and Conservation (Oetting et al. 2006). The F-TRAC analysis is rerun once a year to incorporate the most recent data layer updates, recent acquisitions, and recent losses of natural resources to development.

F-TRAC provides decision-makers with a concise measure of which projects contribute most efficiently to natural resource acquisition goals and illustrates trade-offs that come with prioritizing some sites or resource types over others.

LANDSCAPE CONNECTIVITY: INCORPORATING SPATIAL CONTEXT INTO PROJECT EVALUATIONS

As noted, the original F-TRAC analysis did not include a resource feature for the Florida Ecological Greenways Network (FEGN) or any measure of connectivity generally. A shortcoming of the standard Marxan platform is that it lacks the ability to incorporate spatially explicit connectivity into the analysis, although a more recent tool, Marxan Connect, can incorporate connectivity (Daigle et al. 2020). The standard Marxan platform does include a boundary length modifier feature, which clusters sites more than in a standard optimization, but it is not strategic and does not address important connectivity features, such as differing connectivity priorities for different areas or species or the ability to credit a portfolio for completing any given connection. To address these issues, the team developed the "Greenways for F-TRAC" analysis, an innovative approach to connectivity assessment made possible because the F-TRAC analysis is rerun each year. This analysis models additional priorities within Priority 1 FEGN corridors (a.k.a. "critical linkages") based primarily on distance to conservation hubs (large aggregations of conservation lands), and distance to a corridor's edge. Land cover is also considered, along with elevation above sea level for coastal corridors. As a result, this "cost-distance" model prioritizes building corridors outward from existing conservation lands, interior over edges of corridors, on natural land cover when possible, and above 2 meters of elevation (to address potential sea-level rise effects; fig. 3.5). This method is similar to the "connectivity as conservation feature" approach in Marxan Connect. However, to avoid the limitations of static connectivity modeling, the model is rerun each year to account for new conservation lands and development, and the priorities are extended further along the corridor and shifted as necessary (FNAI 2023c).

With passage of the Florida Wildlife Corridor Act in 2021 (see chapter 7), additional funding and attention were focused on the Florida Wildlife Corridor, which consists of the top three priorities of the FEGN. Accordingly, the FDEP funded a collaboration between the University of Florida Center for Landscape Conservation Planning (UF CLCP) and FNAI to revise, update, and maintain the Florida Ecological Greenways Network to inform acquisition and planning

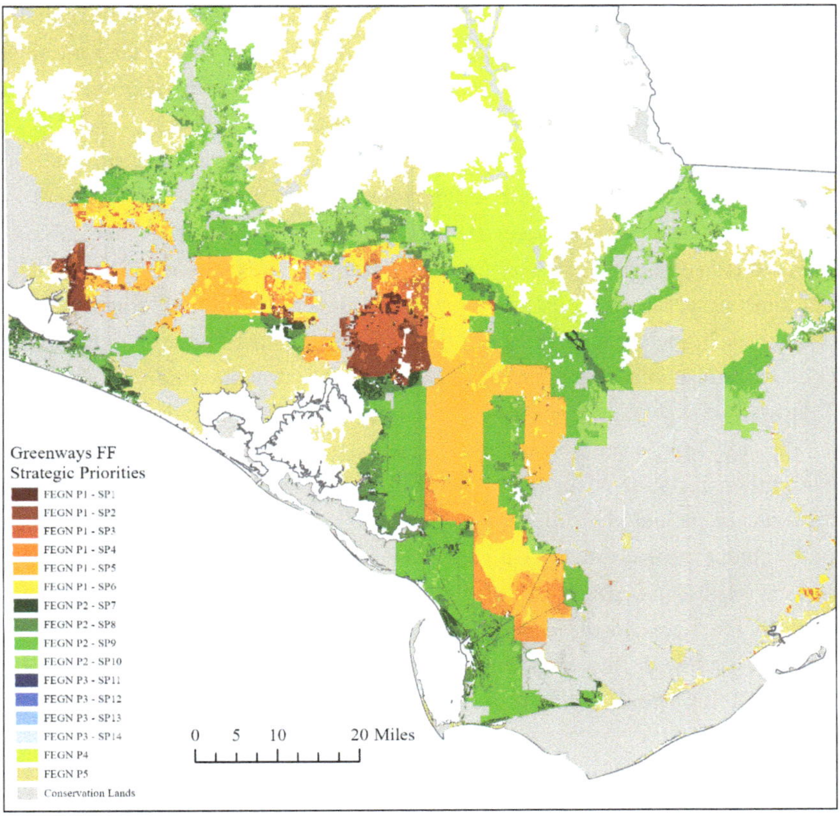

FIGURE 3.5. Detail of Greenways Florida Forever Strategic Priorities map (a.k.a. "Greenways for F-TRAC") in North Florida. While individual strategic priorities classes cannot be as easily distinguished, it illustrates the range of priorities modeled within the underlying FEGN boundaries.

efforts. Both the base boundary and prioritization of the FEGN were updated based on current land cover and conservation lands as well as updates to underlying resource priorities.

As part of this effort, the team expanded on the Greenways for F-TRAC analysis to create the Florida Forever Strategic Priorities layer, a further prioritization of the full Wildlife Corridor to inform Florida Forever project evaluations. This included the original F-TRAC "cost-distance" analysis, but added additional factors: (1) strategic corridors for landscape-scale focal species—key connectivity "bottlenecks" for Florida black bear, Florida panther, and eastern indigo snake

(*Drymarchon couperi*); (2) coastal connectivity—an analysis of connectivity of coastline through natural and seminatural landscapes to at least 3 meters above sea level; (3) threat of land-use conversion—areas overlapping 2040 and 2070 development projections developed by a collaboration between 1000 Friends of Florida and the UF CLCP (1000 FOF 2023); and (4) Fragmentation Index—a neighborhood analysis of the relative proportion of natural and seminatural land uses surrounding any location, where areas with high percentages of natural and low-intensity land uses receive the highest priority scores. The goal of this analysis was to highlight specific areas within the Florida Wildlife Corridor that are the highest priorities for acquisition to maintain landscape connectivity. Further analysis of bottlenecks and threats is underway at the time of publication. This analysis now plays a prominent role in evaluating Florida Forever projects.

The Florida Forever Conservation Needs Assessment and F-TRAC enable a comprehensive suite of assessments for evaluation of land acquisition efforts. Projects are evaluated for direct, on-site single-resource values (species habitat, water quality protection, etc.) as well as for how they fit into a larger landscape context (landscapes single-resource evaluation). Projects are also evaluated holistically for their contribution to the efficient conservation of a broad range of resources statewide, including landscape connectivity (F-TRAC). And progress toward all these goals can be tracked transparently for program accountability.

FLORIDA FOREVER ANALYSIS PRODUCTS: COMMUNICATING COMPLEXITY

With so much data and analyses available, an additional challenge has been distilling complex information into meaningful, clear reports and tools to inform decision-makers. FNAI worked with the Florida Forever Expert Working Group to develop several reporting products. An early report designed to display the full range of individual project scores for both single-resource evaluations and F-TRAC is the Project Comparative Analysis Table (commonly referred to as the "bubble sheet" due to its inspiration from Consumer Reports bubble-style product scores). This report allows users to compare at-a-glance scores for each resource category across the full list of projects; symbols were chosen so that higher scores would visually pop in the context of surrounding scores. For the first several years of Florida Forever, this was the primary guide to project natural resource values provided to ARC in advance of their annual project ranking meeting.

Both FNAI and UF CLCP have also developed a series of web tools to inform land acquisition. Web mapping services have become an increasingly valuable tool for sharing complex GIS data with nontechnical users. FNAI's primary acquisition-related web map is the Florida Forever Conservation Data Viewer, which includes the FFCNA data layers along with conservation lands, Florida Forever project boundaries, and the Florida Wildlife Corridor (FNAI 2024). Users can quickly see which resources overlap with any given project. UF CLCP has developed a map viewer and dashboard for the Florida Ecological Greenways Network (UF CLCP 2024). The dashboard tracks acres of the FEGN protected on conservation lands as well as the percent of the network threatened by projected development based on UF's latest 2040/2070 development project. FNAI was recently contracted by the Live Wildly Foundation to track all land acquisitions at the parcel level (previously this was only done for Florida Forever program acquisitions) and share progress on an online dashboard (FNAI 2024b).

In the 2010s, Greg Brock at the FDEP Division of State Lands developed a comprehensive Excel spreadsheet that collected available data for nearly every measure and criterion outlined in the Florida Forever Act. The FFCNA measures data were central, but the spreadsheet included data for a wide array of criteria ranging from recreational attributes to storm surge to military base buffers. This spreadsheet was provided to ARC members in advance of the annual project ranking meeting and included a scoring system that allowed members to weight measures and criteria according to their preferences. In 2019 following Brock's retirement, FDEP transferred maintenance of the spreadsheet to FNAI. With significant revisions, the spreadsheet is now termed the ARC Ranking Tool and is still provided to council members annually in advance of their December ranking meeting.

CLIP

While not designed to inform land acquisition directly, the Critical Lands and Waters Identification Project, or CLIP, is another statewide planning analysis worthy of noting here. In 2005 the Florida legislature established the Century Commission for a Sustainable Florida, whose charge was to envision Florida over the next 50 years and make recommendations to the governor and legislature regarding sustainable growth. One need identified by the Century Commission was a definitive map of the critical natural resources in Florida that should be considered in any future planning efforts. As conservation planning experts were meeting to address this issue, the Florida Fish and Wildlife Conservation Commission (FWC) undertook a similar effort known as the Cooperative

Conservation Blueprint. FNAI, UF CLCP, and FWC staff working on the two projects pooled their efforts to create the Critical Lands and Waters Identification Project, which consists of a series of statewide GIS data layers depicting a variety of natural resource planning information.

The CLIP core data layers are based largely on the Florida Forever Conservation Needs Assessment measures data. The CLIP team developed some additional data, but what sets the database apart is a series of resource prioritizations: aggregated priority models for Biodiversity, Landscapes, and Surface Water; and a final CLIP Priorities model that combines all three. Working with a technical advisory group and other experts, the CLIP team carefully considered how "critical" resources would be defined and portrayed on maps. Ultimately, the group decided on not a single "critical" designation, but a series of five conservation priorities across the state. While not as focused on specific acquisition targets as the Florida Forever analyses described above, CLIP provides a broad definition of conservation priorities that inform a variety of planning efforts across the state and has been used in state, regional, and local conservation and land-use planning (Oetting et al. 2016). Both UF CLCP and FNAI continue to collaborate on CLIP and anticipate a version 5.0 data update in 2026 that will include updated core data layers, resource prioritizations, and new categories addressing storm protection priorities and surface water restoration.

Rural and Family Lands Protection Program

The Rural and Family Lands Protection Act was passed alongside Florida Forever in 2001, but the program gained prominence more recently as more attention was focused on preserving farmland and protecting the Florida Wildlife Corridor. The Rural and Family Lands Protection Program (RFLPP) is administered by the Florida Department of Agriculture and Consumer Services (FDACS) and is an entirely less-than-fee conservation easement program designed to preserve working agricultural landscapes. For most of the program's duration, projects were evaluated internally by FDACS staff. In 2023, FDACS contracted with UF CLCP, FNAI, and Common Ground Ecology Inc. to develop a more comprehensive assessment of projects. The team developed a three-part assessment: GIS scoring, expert review and narrative, and field visit reporting. Common Ground staff visited each site, confirmed agricultural operations, ground-truthed land cover, and completed a standard quantitative assessment. FNAI staff reviewed each site on-screen with a variety of GIS data and completed a brief narrative description of salient features. GIS scoring combined two broad categories: (1) agricultural and economic value and sustainability and (2) natural resource value, landscape context, and condition. The natural resource data included many of the layers common to the FFCNA and CLIP, with some additional landscape context measures and fire history.

The agricultural/economic layers include several new analyses developed by UF CLCP for this project, created using a combination of soils, land cover, landscape, social, and economic data related to agricultural resource priority. These include an agriculture landscape priorities layer that assesses connectivity within the agricultural and natural landscape, soils-based agricultural productivity suitability, and threat of conversion of agricultural lands to development. FDACS staff relied heavily on this assessment to conduct a final ranking of all 273 projects submitted for the 2023 application cycle. The research team is contracted again to continue to review, refine, and update the evaluation data and process to assist the ranking process. Furthermore, the RFLPP program and the rankings merging natural resource and agricultural data highlight the importance of the relationship between natural resource and green infrastructure values and the economic and quality of life values of agricultural lands, as much of Florida's remaining landscape-scale priority areas for conservation protection are low-intensity and other agricultural lands, including ranchlands, silviculture, and citrus. Conserving these lands benefits not only natural resources but also a rural quality of life and economic system that is being eroded by the steady expansion of suburban and exurban development.

Summary: The Role of a Landscape Perspective in Conservation Land Acquisition in Florida

For more than 30 years, Florida has been a nationwide leader in both land acquisition for conservation and the planning analyses and tools needed to guide efficient conservation. Throughout this time, landscape context and connectivity have been a major focus of conservation efforts. This wealth of experience in planning and acquisition has produced some important lessons. First, landscape context and connectivity require a different level of analysis from many other natural resource features. It is not sufficient to assess landscape value within a site's boundaries; we must consider the spatial context of the site—what and where it connects to (Forman 1987). This requires a higher level of strategic assessment than may be typical of site evaluations and representation tools such as Marxan without complementary and strategic assessment of ecological landscape connectivity. A project may not feature exceptional resource priorities on-site but nevertheless may form a key connection between surrounding priorities. Conversely, planners must be cautious not to use landscape value as an excuse for acquisitions of dubious merit. Landscape value derives from a strategic assessment of a large study area and an individual site's ecological function within it; not all pastures or pine plantations have high landscape value. Nevertheless, our landscape focus has contributed to tremendous progress in conserving a functionally connected network of natural and seminatural lands across Florida and today is guiding ambitious efforts to conserve the Florida Wildlife Corridor.

Two land acquisition projects serve to highlight the value of a landscape perspective in conservation. Located in Walton County in the northwest Panhandle, Seven Runs Creek Conservation Easement forms a critical connection between Eglin Air Force Base to the west and Choctawhatchee River Water Management Area to the east. Totaling about 8,400 ha (20,800 ac), Seven Runs Creek was the vision of M. C. Davis, a private landowner and conservationist, who closed a conservation easement with the FDEP in 2013. The site was historically dominated by longleaf pine sandhill but before the time of easement purchase had been converted to row crops and pine plantation (FNAI 2008). Much of the natural resource value of the site had been degraded, but it was located within an important east–west corridor for Florida black bear and other species, identified in the Florida Ecological Greenways Network since its original delineation. The Davis Estate is now working to restore the natural communities on Seven Runs Creek to significantly increase its intrinsic ecological value, functional ecological connectivity, and ecosystem service provision.

In southwest Florida, the Caloosahatchee Ecoscape Florida Forever project consists of multiple ownerships in Glades and Hendry Counties. This project was originally approved under the Preservation 2000 program in 1999 and carried forward to Florida Forever. The project features a mosaic of natural forests and wetlands, field crops, and pastures. The primary value of the project is the critical linkage it provides for the imperiled Florida panther population to expand northward across the Caloosahatchee River. Currently about 46% of the project is acquired as less-than-fee conservation easements, allowing agricultural activities to continue while ensuring a migration path for the panther. Taken together,

Seven Runs Creek and Caloosahatchee Ecoscape provide tangible examples of the importance of a landscape perspective in conservation and the dedication of private landowners and land acquisition staff in bringing complex projects to completion. A comprehensive, integrated network of functioning landscapes across Florida from the Panhandle to the Keys remains a tall order, but with a history of robust land acquisition programs and innovative conservation planning (combined with other conservation tools such as private landowner incentive programs), it is an achievable goal.

Literature Cited

1000 Friends of Florida (1000 FOF). 2023. Florida's rising seas: Mapping our future. Summary https://1000fof.org/sealevel2040/#. Accessed April 2024.

Ball, I. R., H. P. Possingham, and M. E. Watts. 2009. Marxan and relatives: Software for spatial conservation prioritization. Pages 185–210 in A. Moilanen, K. A. Wilson, and H. P. Possingham, editors. Spatial Conservation Prioritization: Quantitative Methods and Computational Tools. Oxford University Press, Oxford, UK.

Cox, J., R. Kautz, M. MacLaughlin, and T. Gilbert. 1994. Closing the gaps in Florida's wildlife habitat conservation system: Recommendations to meet minimum conservation goals for declining wildlife species and rare plant and animal communities. Florida Game and Fresh Water Fish Commission, Tallahassee.

Daigle, R. M., A. Metaxas, A. C. Balbar, J. McGowan, E. A. Treml, C. D. Kuempel, H. P. Possingham, and M. Beger. 2020. Operationalizing ecological connectivity in spatial conservation planning with Marxan Connect. Methods in Ecology and Evolution. https://doi.org/10.1111/2041-210X.13349.

Farr, J. A., and O. G. Brock. 2006. Florida's landmark programs for conservation and recreation land acquisition. Sustain 14.

Florida Department of Environmental Protection (FDEP). 2024a. SCORP: Outdoor Recreation in Florida. https://floridadep.gov/parks/florida-scorp-outdoor-recreation-florida. Accessed April 2024.

Florida Department of Environmental Protection (FDEP). 2024b. Florida Forever March 2024 Report. https://floridadep.gov/floridaforever. Accessed June 2024.

Florida Greenways Commission (FGC). 1994. Report to the Governor: Creating a statewide greenways system: For people . . . for wildlife . . . for Florida. Florida Department of Environmental Protection, Tallahassee.

Florida Natural Areas Inventory (FNAI). 2008. Preliminary Assessment of Seven Runs Creek. Report prepared for Florida Department of Environmental Protection, October 2008.

Florida Natural Areas Inventory (FNAI). 2023a. Florida Forever Natural Resource Acquisition Progress Report, November 2023. Report prepared for the Florida Department of Environmental Protection. https://www.fnai.org/conslands/florida-forever. Accessed April 2024.

Florida Natural Areas Inventory (FNAI). 2023b. Florida Forever Conservation Needs Assessment Technical Report, version 5.1. November 2023. Report prepared for the Florida Department of Environmental Protection. https://www.fnai.org/conslands/florida-forever. Accessed April 2024.

Florida Natural Areas Inventory (FNAI). 2023c. Florida Forever Project Ranking Support Analyses Documentation, December 2023. Report prepared for the Florida Department of Environmental Protection. https://www.fnai.org/conslands/florida-forever. Accessed April 2024.

Florida Natural Areas Inventory (FNAI). 2024a. Florida Forever Conservation Data Viewer web mapping service. https://geodata.fnai.org/. Accessed April 2024.

Florida Natural Areas Inventory (FNAI). 2024b. Florida Conservation Lands Acquisition Dashboard. https://www.arcgis.com/apps/dashboards/c642f57e0f144487b28779267dd8d8b6. Accessed July 2024.

Forman, R. T. T. 1987. The ethics of isolation, the spread of disturbance, and landscape ecology. Pages 213–229 in M. G. Turner, editor. Landscape Heterogeneity and Disturbance. Springer-Verlag, New York.

Henderson, C. 2022. Forces of Nature: A History of Florida Land Conservation. University Press of Florida, Gainesville.

Hoctor, T. S., W. L. Allen III, M. H. Carr, P. D. Zwick, E. Huntley, D. J. Smith, D. S. Maehr, R. Buch, and R. Hilsenbeck. 2008. Land corridors in the Southeast USA: Connectivity to protect biodiversity and ecosystem services. Journal of Conservation Planning 4:90–122.

Hoctor, T. S., M. H. Carr, and P. D. Zwick. 2000. Identifying a linked reserve system using a regional landscape approach: The Florida Ecological Network. Conservation Biology 14:984–1000.

Hoctor, T., R. Noss, R. Hilsenbeck, J. Guthrie, and C. Ward. 2015. The history of Florida Wildlife Corridor science and planning efforts. https://floridawildlifecorridor.org/wp-content/uploads/2011/12/FWC_History_11_09_2015.pdf. Accessed April 2024.

Jenkins, R. E. 1978. Heritage classification: The elements of ecological diversity. Nature Conservancy News 38(1): 24–25, 30.

Jenkins, R. E. 1985. Information methods: Why the heritage programs work. Nature Conservancy News 35(6):21–23.

Knight, G. R., J. Oetting, and L. Cross, editors. 2011. Atlas of Florida's Natural Heritage. Florida State University, Tallahassee.

Oetting, J., T. Hoctor, and M. Volk. 2016. Critical Lands and Waters Identification Project (CLIP): Version 4.0 Technical Report. Report to the U.S. Fish and Wildlife Service. Accessed April 2024. https://www.fnai.org/services/clip.

Oetting, J. B., A. L. Knight, and G. R. Knight. 2006. Systematic reserve design as a dynamic process: F-TRAC and the Florida Forever program. Biological Conservation 128:37–46.

University of Florida Center for Landscape Conservation Planning (UF CLCP). 2024. Florida Ecological Greenways Network map viewer and dashboard. https://conservation.dcp.ufl.edu/fegn/. Accessed April 2024.

4

Statewide Land-Use Planning and Growth Modeling

Margaret (Peggy) Carr, Richard Doty, Daniel Farrah, Jason Teisinger, Liz Thompson, Michael Volk, and Paul Zwick

Introduction—Why Spatially Model Population Growth for Landscape Conservation Planning?

It has been well documented and reported that Florida is a state of rapid population growth (Perry et al. 2022; Rayer et al. 2024; Wilder et al. 2024). Though outpaced by Nevada for many years in terms of percentage growth, in 2022 Florida earned the title of fastest growing state (Perry et al. 2022). The population more than doubled between 1980 and 2020, increasing from 9,746,961 to 21,538,187 (fig. 4.1; US Census Bureau 2021; Rayer et al. 2021). As these numbers catapulted, it became increasingly obvious to all involved in landscape conservation planning that understanding the impacts of population growth on patterns of land use was imperative. Knowing that population was on the increase was in and of itself important, but understanding how that population was being distributed across the state came to be an even more important consideration. The need for spatial population modeling was clear.

Spatial population models have been used for many purposes, including transportation modeling and water supply planning and permitting. In the context of conservation planning, three primary purposes of population modeling have been

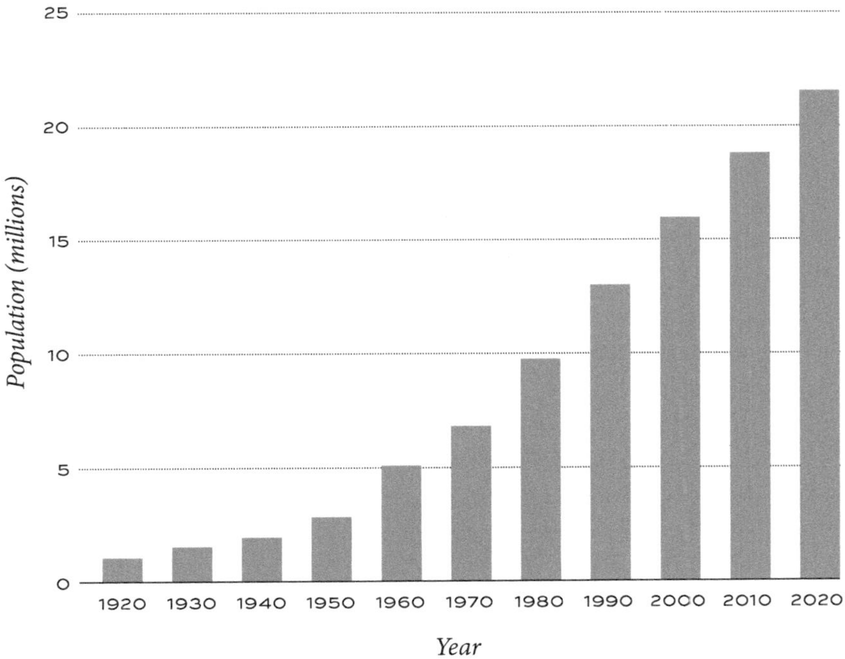

FIGURE 4.1. Decennial population growth in Florida, 1920–2020. Data from US Census Bureau.

1. to graphically demonstrate growth trends in order to motivate the public and state officials to invest in permanent protection of large land areas;
2. to assist in prioritizing the expenditure of time and money on land conservation; and
3. to nudge local governments toward the adoption of land-use policies that yield more compact communities in order to protect working landscapes and natural areas.

The first state-sanctioned landscape conservation plan was adopted by the Florida Greenways Commission in 1998, and it came in the form of the Florida Ecological Greenways Network (FEGN), part of the *Five-Year Implementation Plan for the Florida Greenways and Trails System* (see chapter 7, this book). This followed two earlier efforts of note, the statewide corridor network map produced by Reed Noss in the 1980s (see chapter 2, this book) and the results of a workshop held in 1991 coordinated by The Nature Conservancy that used a variety of experts to identify conservation priorities (Hoctor et al. 2000; see chapter 3, this book).

In the 1998 plan, the total land area identified as having the potential to be included in the FEGN was just shy of 7.7 million ha (19 million ac), representing roughly 47% of the state (Carr et al. 1999). A number of large landowners expressed concern about having their lands included in this designation for fear that inclusion might lead to regulatory constraints. In response, the mapping and associated spatial data clearly acknowledged that the mapping effort should not be used for regulatory purposes. Nonetheless, this concern slowed the effort to permanently protect areas within the FEGN, which led directly to the idea of spatially modeling population growth to demonstrate what might happen should land protection stall and existing land-use policies continue, coupled with an ever-growing population. 1000 Friends of Florida, under the leadership of Board Member Tim Jackson, was one nongovernmental organization that recognized very early on just how critical this effort would become.

The GIS modeling process used to create the first version of the FEGN included a long list of criteria, most of which were indicative of suitability for conservation; but there were a few criteria that addressed lack of suitability. At that time, no statewide population distribution modeling had been done, so surrogate data were used to identify less suitable areas for development. These included the presence of and proximity to existing developed lands and areas of high road density (Hoctor et al. 2000). At least the static assessment of areas of low suitability was considered using these criteria; however, there was no way at that time to anticipate what future patterns of population distribution would yield. Fortunately, however, more robust mapping techniques, enhanced population and other land-use datasets, and steadily increasing computational power were on the way. These developments would contribute to more sophisticated conservation prioritization.

Using Spatial Growth Models in Landscape Conservation Planning

In the late 1990s planners and scientists in Florida and elsewhere began to make significant progress in the modeling of future population distribution. Milestones are described here in chronological order beginning with a 1997 effort and concluding with one completed in 2024. Each description includes a commentary on its relevance to landscape conservation planning.

The need for these products as part of landscape conservation planning was formally acknowledged in 1998 when researchers at the University of Florida were asked to assist with prioritizing lands identified as part of the FEGN (Carr et al. 1999; Hoctor et al. 2001). Through the exercise, six priority classes were identified, but population distribution models were not available and so were not used as part of the effort. However, the 2001 report went on to note, "Another step in the prioritization process that should be considered is to compare the high priority results with development pressure. . . . Protection efforts should concentrate primarily on areas within the high priority results that are also in areas most threatened by development" (Hoctor et al. 2001).

PREDICTING SPATIAL DISTRIBUTION OF FUTURE POPULATION FOR ORANGE COUNTY, FLORIDA (1997)

University of Florida master's student Richard Doty undertook the design of an early GIS model to predict future population growth in Orange County, Florida, for his 1997 thesis project. Orange County partially sits within the St. Johns River Water Management District, and population predictions were needed by the District to plan for future water demand (Doty 1997).

The 1990 population from the US Census was distributed into the square mile sections of Orange County based on residential land-use data derived from the Department of Revenue parcel database. Four different population projection methodologies (linear, exponential, share of growth, and shifted share of growth) were used to project the 2020 population. An average of the four results was then normalized with the Bureau of Economic and Business Research's (BEBR's) projected population increase for the same period to provide the population projections used for each section (Doty 1997).

Nondevelopable lands were identified and made off-limits for the distribution of new population. These included road rights-of-way, water bodies, conservation lands, and wetlands. Then a suitability analysis was developed from features that are known to attract growth, including major roads, residential areas, commercial areas, and water bodies. A maximum population density was determined for each section based on a statistical analysis of its 1990 population density. If the projected population growth exceeded the section's available capacity, the "lost" growth was stored and later distributed to sections with available capacity and high-growth suitability. The model was run iteratively every five years beginning in 1990 and ending in 2020 (Doty 1997). The final results were summarized by the boundaries of the water utilities that service each square mile section, providing each utility with a forecast of the change in population it must serve.

This model was not used for landscape conservation planning, but it is an early example of a spatial population distribution exercise used for resource planning. It could just as easily have been used to predict the likelihood of population growth in areas of ecological significance, suggesting the need for conservation action.

PREDICTING STATEWIDE GROWTH POTENTIAL (2002)

University of Florida master's student Jason Teisinger undertook the design of a GIS model to predict future statewide growth potential for his 2002 thesis project. This was the first effort of which we are aware used to predict the potential spatial distribution of future population for the entire state of Florida. Its primary purpose was to "identify areas with greatest potential for conversion from non-urban to urban land uses" (Teisinger 2002).

The US Census Bureau estimate of the 1995 population for each county in Florida was used to determine the persons/urban acre for each. The BEBR (2001) medium and high projections for 2030 were used to predict areas of conversion from nonurban to urban land uses (Teisinger 2002).

The model was composed of five primary analyses: location, historic change, vacant residential parcels, projected census change, and a nondevelopable mask. For counties with sufficient undeveloped land to accommodate the projected population increase at historic densities, it was assumed all new development would occur on greenfields. For three counties—Broward, Dade, and Pinellas—insufficient greenfield land was available to accommodate all the projected growth, so it was assumed that the density of development would increase to accommodate the population increase. The results yielded a 2030 statewide population allocation (Teisinger 2002).

The 2030 allocation was used to analyze the vulnerability of the priority linkages of the then latest version of the FEGN. This was done by creating three growth potential classes from the allocation—high, medium, and low—and comparing these with three classes of priority linkages: high, medium, and low. The combination produced a representation of the most critical priority linkages. The top five identified were Tosohatchee to Ocala National Forest, Ocala National

Forest to Okefenokee, Big Cypress to Avon Park, Wekiva to Ocala National Forest, and Green Swamp to Withlacoochee to Annutteliga (Teisinger 2002).

FLORIDA 2060 VERSION 1 (2006) AND VERSION 2 (2013)

In 2005 as the reality of Florida's rapid population growth was becoming crystal clear, 1000 Friends of Florida engaged the University of Florida's GeoPlan Center to develop representations of future population distribution to graphically demonstrate the change the state was likely to undergo. Spatial population modeling techniques and supporting hardware had evolved to the point where complex statewide population distribution modeling was finally feasible. The results of the effort were made available in 2006 (Zwick et al. 2006).

Simultaneously, 1000 Friends engaged Georgia Tech's Center for Quality Growth and Regional Development to prepare a study called "A Time for Leadership: Growth Management and Florida 2060," which addressed land-use development and management policies (Ross 2006).

The Florida 2060 Version 1 (2006) effort stimulated considerable interest and concern about the future of development and conservation in the State. However, it only included one population distribution scenario for each of the years 2020, 2040, and 2060. Nevertheless, the results were widely distributed by 1000 Friends of Florida, and the general public began to see what the future might hold. In 2012, Zwick and Carr undertook to model three new 2060 alternative population distribution scenarios (2060 Version 2) for the University of Florida's Center for Landscape Conservation Planning, each based on a different set of assumptions about land-use policies and potential sea-level rise (Zwick et al. 2013).

The county population projections used in 2060 Version 2 were derived from the BEBR 2012 medium projections that extended out to 2040. A straight-line extension of the change from 2020 to 2040 was used to estimate the population change through 2060. The Version 2 projection was nearly six million less than in Version 1. This was likely the result of the 2008 recession being taken into consideration by BEBR, as it slowed Florida's population growth (Zwick et al. 2013).

In Florida 2060 Version 1, the three scenarios developed for 2020, 2040, and 2060 were based on the same assumptions with the only variable being population. In Version 2, three scenarios were modeled—trend, sea-level rise 1, and sea-level rise 2—each based on the same projected population increase but with a different set of other assumptions.

The following assumptions were used for all three scenarios:

1. No new conservation lands would be set aside for future projections because it would be purely speculative to identify them.
2. Existing conservation lands were a combination of fee simple and less-than-fee simple properties under permanent protection held by government entities and nongovernmental organizations.
3. Analysis of nine weighted criteria determined a range of suitability of the lands available in each county for future development. This varied only slightly from the suitability analysis used in Version 1.

The most significant difference in the allocation process between Version 1 and Version 2 was the use of a redevelopment variable. This was added because it was becoming obvious that not all population growth was being accommodated on greenfields but that some was being accommodated on previously developed lands that were being redeveloped at higher densities. The more urban counties had the higher redevelopment values, as one might predict. These values were then used to reduce the amount of greenfield land needed to accommodate the future population (Zwick et al. 2013).

In the trend scenario the projected 2060 population increase over 2010 was allocated for each county. A county-by-county calculation of gross urban density determined the people per acre to be allocated into the most suitable available lands. The densities ranged from a high of 14.13 people/acre in Dade County to a low of 0.42 people/acre in Levy County (Zwick et al. 2013).

In the sea-level rise 1 scenario a calculation was made of the people to be displaced by 1 m of sea-level rise plus a buffer of lands that would be inundated by an additional 0.5 m of sea-level rise by intersecting 2010 census data with a representation of the area included in a 1.5 m sea-level rise. This displaced population was added to the 2060 population projection used for the trend scenario. It was assumed all displaced individuals would relocate within the state. A statewide representation of greenfield lands suitable for development was used to distribute the projected population increase plus the displaced population (Zwick et al. 2013).

In sea-level rise 2, the population distribution resulting from the trend scenario was used as the starting point. Then the same displaced population used for sea-level rise 1 was allocated using the same statewide representation of suitability used in sea-level rise 1. Again, it was assumed all displaced individuals would relocate within the state (Zwick et al. 2013).

Contrast was evident among the three scenarios, but in fact none were used in the *Sea Level Rise Biodiversity Assessment* undertaken by the University of Central Florida, Florida Natural Areas Inventory, and University of Florida's Center for Landscape Conservation Planning. Instead, Florida 2060 Version 2 stimulated substantive discussion about modeling modifications that would produce more nuanced and realistic population distribution scenarios that included sea-level rise, for use in landscape conservation planning (Hoctor et al. 2014).

PENINSULAR FLORIDA LANDSCAPE CONSERVATION COOPERATIVE: LANDSCAPE CONSERVATION AND CLIMATE CHANGE SCENARIOS FOR THE STATE OF FLORIDA (2014)

In 2014, GeoAdaptive, LLC, published an update to a 2012 study commissioned by the Peninsular Florida Landscape Conservation Cooperative (PFLCC) that extended the geographic range of the scenarios from the original Peninsular Florida cooperative boundaries to include the entire state of Florida. The goal was to develop future development scenarios that utilized sea-level rise, financial resources, population growth, conservation strategies, and land-use and water assumptions, policies, and regulations (GeoAdaptive 2014).

The study consisted of three future development scenarios, each using a projected population of 31 million residents in the state by 2060 based on county-level projections from BEBR (GeoAdaptive 2014). The population graph included in the report suggests the population figure was generated using a linear extrapolation of 2040 BEBR population projections available in 2014.

The three scenarios developed were a trend or business-as-usual scenario and two conservation scenarios. Each was driven by six assumptions: acquisition rate for conservation, conservation strategy and prioritization, minimum lot size for conservation, urbanization/planning strategy, projected future population (31,000,000), and projected sea-level rise (1 m by 2100). The latter two assumptions remained constant for all three scenarios. For the trend scenario, the acquisition rate for conservation was 12,141 ha/yr (30,000 ac/yr) with the majority protected in fee simple with no minimum conservation parcel size. The urbanization/planning assumption was continued low-density greenfield development at existing densities. Scenario 2 assumed an acquisition rate for conservation of 18,211 ha/yr (45,000 ac/yr) with the vast majority protected by easement with no minimum conservation parcel size. The urbanization/planning assumption was proactive green infrastructure with increased redevelopment and densities. Scenario 3 assumed an acquisition rate for conservation of 23,067 ha/yr (57,000 ac/yr) with the vast majority protected by easement with a minimum conservation parcel size of 2 ha (5 ac) for fee simple and 40.5 ha (100 ac) for easements. The urbanization/planning assumption was the same as for scenario 2: proactive green infrastructure with increased redevelopment and densities (GeoAdaptive 2014).

A cohesive, statewide network of conservation lands was found to be possible in all three scenarios with reasonable levels of financial support based on historical expenditures. The trend scenario was found to largely accommodate Florida Forever goals by 2040 but with gaps remaining in connectivity of conservation areas. The conservation scenarios as expected would lead to large increases in conservation relying heavily on easements. The primary lesson gained from these models was the need to develop strong and coordinated institutions to manage future growth and conservation lands (GeoAdaptive 2014).

FLORIDA 2070 (2016)

In 2015, 1000 Friends of Florida engaged UF's GeoPlan Center to prepare an update that would use the newest data available and would include three scenarios: baseline 2010 that represented the actual distribution of development and conservation lands in 2010 (to serve as a basis for comparison), trend 2070 that depicted the possible distribution of population in 2070 should existing land-use policies and development densities remain the same, and alternative 2070 that accommodated the 2070 projected population but with a more compact pattern of development and more land conservation (Carr et al. 2016).

The population projection used for this effort was the 2015 medium-range projection prepared by the BEBR that extended to the year 2040. A straight-line extrapolation was used to project 2070 population estimates for each county. The total 2070 population for the state was 33,721,828, approximately 3,000,000 less than the projection used in the Florida 2060 Version 2 (Carr et al. 2016).

Table 4.1 compares the major modeling assumptions used for the trend and alternative scenarios. The different assumptions resulted in markedly different results. Table 4.2 quantifies the results in comparison with the baseline.

The single most important finding of the study was that even modest increases in development densities can result in substantial savings of conservation and agricultural lands. The contrast in the quantity and type of lands included in

TABLE 4.1. Comparison of major modeling assumptions for 2070 Trend and 2070 Alternative, from Carr et al. (2016)

	2070 TREND	2070 ALTERNATIVE
Redevelopment areas	No new population accommodated within existing urban areas	In all counties, some of the 2070 population growth is accommodated through an increase in density
Development densities	New population allocated at each county's 2010 development density	After removing the redevelopment population, the remaining new population is allocated at a development density 20% greater than that used for the trend
Protected lands	2016 Florida Managed Areas included as protected	In addition to 2016 Florida Managed Areas, 2015 Florida Forever project lands, and FEGN Priorities 1 and 2 added to protected status
Agricultural lands	No agricultural lands excluded from population allocation	All irrigated agricultural lands (2035) on good soils (USDA/NRCS Capability Rating Excellent–Marginal) from FSAID, 2015 excluded from population allocation

TABLE 4.2. An acreage comparison of Florida 2070 alternative population allocation scenarios

	BASELINE	% OF LAND	TREND	% OF LAND	ALTERNATIVE	% OF LAND
Developed	6,275,000	18.17	11,648,000	33.72	9,777,000	28.30
Protected (excluding agriculture)	9,269,000	26.83	9,525,000	27.57	13,339,000	38.62
Protected agriculture	924,000	2.67	1,106,000	3.20	3,120,000	9.03
Agriculture (croplands, livestock, aquaculture)	7,586,000	21.96	5,422,000	15.70	4,513,000	13.06
Other (mining, timber, etc.)	10,489,000	30.37	6,842,000	19.81	3,794,000	10.98
Total state land acres	34,543,000	100.00	34,543,000	100.00	34,543,000	100.00
Acres of open water	1,795,000					
Total Florida acreage	36,338,000					

permanent protection in the trend and alternative scenarios demonstrated how land development policies coupled with land conservation programs can directly affect landscape conservation.

AMERICAN FARMLAND TRUST *FARMS UNDER THREAT 2040: CHOOSING AN ABUNDANT FUTURE* (2022)

In 1993 American Farmland Trust (AFT) developed the *Farming on the Edge* report to understand the geographic relationship between land development pressure and high-quality farmland. A subsequent study in 2018, *Farms under Threat: The State of the States,* was undertaken to document the extent, location, diversity, quality, trends, and threats for agricultural land (American Farmland Trust 2024). A second iteration, *Farms under Threat 2040: Choosing an Abundant Future* (Hunter et al. 2022), modeled the impacts of future urban development and climate change on agricultural lands by 2040. Details for this last effort are also documented in a publication in *Land* entitled "U.S. Farmland under Threat of Urbanization: Future Development Scenarios to 2040" (Xie 2023).

The study used county population estimates from the US Census Bureau for the years 2001 and 2016 along with statewide estimates to determine projected population growth between 2016 and 2040. Projected growth was divided into two types, low-density residential (LDR) and urban and highly developed (UHD) (Xie 2023).

The study was based on several assumptions:

1. Sea level is projected to rise by 19 cm (7.5 in) by 2040 (Xie 2023).
2. Lands unlikely to be developed were removed from potential future growth, including existing urban, protected agricultural lands, lands in the Protected Areas Database of the US, federal lands, and residential forest areas (Xie 2023).
3. The study explored three scenarios: business-as-usual, runaway sprawl, and better-built cities (Hunter et al. 2022) with varying development densities.

The business-as-usual scenario modeled future development as if it followed current patterns (trend scenario). The runaway sprawl scenario analyzed the effects of future urban growth when low-density conversion rates are increased by 50% over the historical rate. The better-built cities scenario focused on reduction of sprawl by modeling the impacts of a 25% reduction of UHD lands and 50% reduction of LDR lands compared to the business-as-usual scenario (Hunter et al. 2022).

One compelling aspect of this effort was the validation of the modeling assumptions through an evaluation of projected growth between 2001 and 2016 and actual growth. Results confirmed a high degree of consistency between the two.

If development continues on its current trajectory as modeled in the business-as-usual scenario, more than 7.4 million ha (18.4 million ac) of farmland and ranchland will be lost or degraded by 2040. Two-thirds of new development will be low-density residential, smaller farms will be highly impacted, and the most productive farmland and ranchland will be 50% more likely to be converted. The runaway sprawl scenario would result in more than 2.4 million ha (6 million ac) more land converted compared to the business-as-usual scenario (Hunter et al. 2022). The better-built cities scenario demonstrated that with smart planning, development can be accommodated while preserving three million ha (7.5 million ac) more land than in the business-as-usual scenario (Hunter et al. 2022).

The focus of this effort was the preservation of farmland, but as has been demonstrated in Florida and other places, protected working landscapes have the potential to benefit conservation as well as the rural human population. The study was made available on AFT's website, which includes a wide range of information to assist in the preservation of farmland. A webinar series for over half the states in the country was used to disseminate the findings.

SEA LEVEL 2040/2070 (2023)

The fourth iteration of Florida population distribution modeling completed at the University of Florida was developed as a response to a need to include the impacts of sea-level rise on future development scenarios and to provide updated projections using newer data. To emphasize the importance of the inclusion of sea-level rise, the project was named Sea Level 2040 and Sea Level 2070. This study largely followed the methodology of the previous studies while using newer data and with some updates to the methodology. A follow-up study based on the Sea Level 2040/2070 study, Agriculture 2040/2070, was developed to explore the impacts from future development on agricultural lands in the state. As with previous versions of this study, 1000 Friends of Florida developed publicly available reports and facilitated public outreach efforts to disseminate the study's findings.

The Sea Level 2070 study used the same population projections as the Florida

TABLE 4.3. Population projections used for Sea Level 2040/2070 (UF CLCP 2023a, 2023b)

PERIOD	STARTING POPULATION	ENDING POPULATION	PERCENT INCREASE	ADDITIONAL POPULATION TO BE ALLOCATED
2019–2040	21,477,737	26,406,000	23	4,928,263
2019–2070	21,477,737	33,721,828	57	12,244,091

2070 study (2016). Using the same population projections made it possible to see the impacts of sea-level rise on future development, both in coastal and inland areas impacted by population in-migration (UF CLCP 2023a, 2023b).

Sea Level 2040 used the 2021 BEBR medium population projections. These updated projections (table 4.3) were deemed appropriate due to the projection date (2040 vs. 2070) for the scenario and because there was no need for data parity with a previous study (UF CLCP 2023a, 2023b).

The 2040 and 2070 scenarios were based on three assumptions:

1. Sea level will rise 0.25 m (9.8 in) by 2040 and 0.9 m (35.4 in) by 2070.
2. For both Sea Level 2040 and 2070, residents on land lost to sea-level rise will relocate, with half relocating nearby to current employment and social networks and half moving out of state.
3. Two future development scenarios were used for both Sea Level 2040 and 2070 as described below.

The two scenarios were a sprawl scenario and a conservation scenario (analogous to the prior trend and alternative scenarios completed in 2016). The sprawl scenario assumed that current development patterns and densities would continue through 2070. The conservation scenario accommodated the same population growth as the sprawl scenario while assuming higher-density development (2040 and 2070) and redevelopment (2070) and also allowing the state's priority natural lands to remain undeveloped. Priority natural lands included the top three priorities of the FEGN, also known as the Florida Wildlife Corridor, and 2023 approved projects in Florida's land acquisition program, Florida Forever. The Sea Level 2070 study used the same gross development density (GDD) calculations as those in the 2016 model. The Sea Level 2040 study used 2019 data to determine the GDD (UF CLCP 2023a, 2023b).

Florida is projected to have 12.2 million more residents by 2070 and could lose 688,000 ha (1.7 million ac) to sea-level rise, with more than 900,000 people needing to relocate. Impacted areas include both coastal and inland counties due to their hydrologic connections with the coast. The overall result is more people moving to the state with less land to accommodate them.

The Sea Level 2070 sprawl scenario projects a loss of 1.4 million ha (3.5 million ac) to development and 688,000 ha (1.7 million ac) to sea-level rise, of which 728,000 ha (1.8 million ac) are currently in agriculture and 769,000 ha (1.9 million ac) in other uses, including silviculture. The conservation 2070 scenario projects 526,000 ha (1.3 million ac) less developed lands, over two million ha (5 million ac) more protected natural lands, and 931,000 ha (2.3 million ac) more protected agricultural lands compared to the sprawl scenario. Development using higher density and incorporating redevelopment can accommodate the projected population increase while protecting priority natural lands.

The results of these models were shared via a public website and webinar series and are being used (in progress as of this writing) to develop a dynamic online public viewer showing areas of potential development threat. As noted earlier, the Sea Level 2040/2070—2023 models also formed the basis for an additional Agriculture 2040/2070—2023 report and website. As part of the Agriculture 2040/2070—2023 work (also ongoing as of this writing), leaders from Florida's agricultural organizations were convened to assist in developing strategies and recommendations for addressing the impacts from future development on agricultural lands and production. The scenarios are also being used as one dataset to prioritize conservation lands based on potential development threat, including to identify agricultural land protection priorities, and as part of ongoing Florida Wildlife Corridor planning. Finally, the methodology for the Sea Level 2040/2070—2023 scenarios is being used as a basis for county-level partnerships to examine impacts at the county scale and discuss priorities and measures for sound future land-use planning.

SUMMATION

This chronological summary of seven population distribution studies reveals many similarities, but also a growing level of nuance and complexity. All were undertaken using raster GIS with data summarized up to either square mile sections or counties. The early models produced only one result each, but after 2013, all included multiple scenarios. For example, the 2022 AFT effort produced three scenarios: business-as-usual, runaway sprawl, and better-built cities (Hunter et al. 2022). Scenarios are important because they demonstrate that changes in land-use policy can result in outcomes that accommodate population growth while saving conservation lands and working landscapes.

Increasingly these products are being used to prioritize conservation of native/seminatural lands and working landscapes based on development threat. For those deep into the details of land conservation, the products are useful because they can provide quantitative measures of potential loss if action is not taken. But these models have also aided the general public in visualizing the future. Land-use changes are often random and incremental; the term "disjointed incrementalism" has been used to describe this phenomenon (Carr et al. 2007). However, by looking decades into the future, incremental changes can be compiled to reveal the long-term result.

Future Efforts as Modeling Techniques and Relevant Data Evolve

The models described in this chapter are primarily based on a simple rules-based approach to suitability modeling. However, GIS-based models, methods, and software, including for future development projections, continue to advance, providing the opportunity for more nuanced, complex, and (potentially) more accurate results.

These potential new approaches include the use of machine learning and artificial intelligence (AI) to incorporate historical development data and develop future scenarios (e.g., see Kim et al. 2022). Models are also still being developed that further incorporate human response and historic migration data to predict potential climate-induced population shifts (e.g., Hauer 2017; Robinson et al. 2020). Several studies have similarly agreed that while individual decisions are not possible to model, there are some common themes to these decisions (Wrathall et al. 2019; Robinson et al. 2020), and the ongoing question is if/how these can be modeled to reflect potential future human responses to climate or other factors. Models are also being designed to incorporate additional data sources such as development permit information (e.g., Hauer 2017; Robinson et al. 2020; CFRPC 2022; Internal Revenue Service data), temporal factors, and remote imagery (Meentemeyer et al. 2013). With ever-increasing computational power, it is also possible that even statewide modeling will become much more resolute, using, for example, data linked to property parcels.

As always, models for future population growth and development are still "only models" that aim to provide future predictions or scenarios. Even the most complex and sophisticated models will be limited in their capacity to predict the future. Nevertheless, with refinements in modeling technology and design, a reasonable goal may be to provide useful results using defensible and clear research methods, furthering land-use planning and research, and helping to proactively address critical land-use and resource questions going forward.

Literature Cited

1000 Friends of Florida and University of Florida Center for Landscape Conservation Planning (UF CLCP). 2023a. Florida's Rising Seas: Mapping Our Future—Sea Level 2040. Prepared for the Florida Department of Agriculture and Consumer Services. https://1000fof.org/sealevel2040/. Accessed September 21, 2024.

1000 Friends of Florida and University of Florida Center for Landscape Conservation Planning (UF CLCP). 2023b. Florida's Rising Seas: Mapping Our Future—Sea Level 2070. Prepared for the Florida Department of Agriculture and Consumer Services. https://1000fof.org/sealevel2040/wp-content/uploads/2023/06/Sea-Level-2070-Report-FINAL.pdf. Accessed September 21, 2024.

American Farmland Trust. 2024. https://www.development2040.farmland.org. Accessed August 7, 2025.

Carr, M. H., P. D. Zwick, T. S. Hoctor, and M. A. Benedict. 1999. Final Report, Phase II, Florida statewide greenways planning project, Department of Landscape Architecture, University of Florida.

Carr, M. N., and P. D. Zwick. 2007. Smart Land Use Analysis: The LUCIS Model, Land Use Conflict Identification Strategy. Esri Press.

Carr, M. H., and P. D. Zwick. 2016. Technical Report: Florida 2070, Mapping Florida's future—alternative patterns of development in 2070. School of Landscape Architecture and Planning, University of Florida. Prepared for the Florida Department of Agriculture and Consumer Services and 1000 Friends of Florida.

Central Florida Regional Planning Council (CFRPC). 2022. Florida Wildlife Corridor Planning and Development Data Pilot Project. Central Florida Regional Planning Council, Bartow.

Doty, R. L. 1997. Predicting spatial distribution of future population for Orange County, Florida. Master's Thesis. University of Florida, Department of Urban and Regional Planning.

Florida Bureau of Economic and Business Research, Florida Estimates of Population 2015, College of Liberal Arts and Sciences, University of Florida, 2015. https://www.bebr.ufl.edu/sites/default/files/Research%20Reports/estimates_2015.pdf. Accessed September 21, 2024.

GeoAdaptive, LLC. 2014. Landscape conservation and climate change scenarios for the state of Florida: A decision support system for strategic conservation. Summary for decision makers. Prepared for the Peninsular Florida Landscape Conservation Cooperative, US Fish and Wildlife Services and US Geological Survey.

Hauer, M. E. 2017. Migration induced by sea-level rise could reshape the US population landscape. Nature Climate Change 7:321–325.

Hoctor, T. S., M. H. Carr, and P. D. Zwick. 2000. Identifying a linked reserve system using a regional landscape approach: The Florida Ecological Network. Conservation Biology 14:984–1000.

Hoctor, T. S., J. Oetting, R. F. Noss, M. Volk, J. Reece. 2014. Predicting and Mitigating the Impacts of Sea Level Rise and Land Use Change on Imperiled Species and Natural Communities. Final Report. Florida Fish and Wildlife Conservation Commission, Tallahassee.

Hoctor, T. S., J. K. Teisinger, M. H. Carr, and P. D. Zwick. 2001. Ecological greenways network prioritization for the state of Florida, a report to the Florida Office of Greenways and Trails. University of Florida, Gainesville.

Hunter, M., A. Sorensen, T. Nogeire-McRae, S. Beck, S. Shutts, and R. Murphy. 2022. Farms under threat 2040: Choosing an abundant future. American Farmland Trust, Washington, DC.

Kays, J. 2023. There's a Map for That. Florida Trend, September 2023.

Kim, Y., A. Safikhani, and E. Tepe. 2022. Machine learning application to spatio-temporal modeling of urban growth. Computers, Environment and Urban Systems. Volume 94. https://doi.org/10.1016/j.compenvurbsys.2022.101801. Accessed September 21, 2024.

Meentemeyer, R. K., W. Tang, M. A. Dorning, J. B. Vogler, N. J. Cunniffe, and D. A. Shoemaker. 2013. FUTURES: Multilevel simulations of emerging urban—rural landscape structure using a stochastic patch-growing algorithm. Annals of the Association of American Geographers 103:785–807.

Perry, M., L. Rogers, and K. Wilder. 2022. https://www.census.gov/library/stories/2022/12/florida-fastest-growing-state.html#:~:text=After%20decades%20of%20rapid%20population%20increase%2C%20Florida%20now,2022%2C%20surpassing%20Idaho%2C%20the%20previous%20year's%20fastest-growing%20state. Accessed September 21, 2024.

Rayer, S., and C. Comfort. 2024. Projections of Florida population by county, 2025–2050, with estimates for 2023. BEBR. Volume 57, Bulletin 198.

Rayer, S., R. L. Doty, S. Roulston-Doty, J. K. Teisinger, and Y. Wang. 2021. Florida

population: Census summary 2020. Bureau of Economic and Business Research, University of Florida.
Robinson, C., B. Dilkina, and J. Moreno-Cruz. 2020. Modeling migration patterns in the USA under sea level rise. PLOS One 15(1): e0227436.
Ross, C. L., K. Leone de Nie, and J. Barringer. 2006. A time for leadership: Growth management and Florida 2060. Center for Quality Growth and Regional Development, Georgia Institute of Technology. Prepared for 1000 Friends of Florida.
Teisinger, J. K. 2002. Predicting statewide growth potential. Master's Thesis. University of Florida, Department of Urban and Regional Planning.
US Census Bureau. 2021. Historical Population Change Data (1910–2020). https://www.census.gov/data/tables/time-series/dec/popchange-data-text.html. Accessed March 27, 2025.
Wilder, K., and P. Mackun. 2024. Florida home to metro areas among top U.S. population gainers. Florida Trend. https://www.floridatrend.com/article/39633/florida-home-to-metro-areas-among-top-us-population-gainers. Accessed September 21, 2024.
Wrathall, D. J., V. Mueller, P. U. Clark, A. Bell, M. Oppenheimer, M. Hauer, S. Kulp, E. Gilmore, H. Adams, R. Kopp, K. Abel, M. Call, J. Chen, A. deSherbinin, E. Fussell, C. Hay, B. Jones, N. Magliocca, E. Marino, and K. Warner. 2019. Meeting the looming policy challenge of sea-level change and human migration. Nature Climate Change 9:898–901.
Xie, Y., M. Hunter, J. P. Suraci, A. Sorensen, T. Nogeire-McRae, R. Murphy, S. Lischka, and T. J. Lark. 2023. U.S. Farmland under threat of urbanization: Future development scenarios to 2040. Land 2023:574.
Zwick, P. D., and M. H. Carr. 2006. Florida 2060: A population distribution scenario for the State of Florida. GeoPlan Center, University of Florida. Prepared for 1000 Friends of Florida.
Zwick, P. D., and M. H. Carr. 2013. Florida 2060 Version 2.0: Three alternative land use scenarios. School of Landscape Architecture and Planning, University of Florida. Prepared for Center for Landscape Conservation Planning, University of Florida.

5

Identifying and Protecting Climate Refugia at Multiple Spatial Scales

Eve Bohnett, Reed F. Noss, Alan Flint, Lorraine Flint, Michael O'Brien, Paul Donsky, Isabella Guttuso Browne, Derya Yesilkusak, and Thomas S. Hoctor

Introduction

Conservation is a long-term endeavor. Conservation planners must address not only the current landscape but also changing conditions over time. Attempts must be made to safeguard the regional biota from a future climate projected to be hostile. Of the options considered for biological adaptation to climate change, two may be most promising in the context of conservation planning (Noss 2001): (1) providing habitat connectivity to allow species to move in response to change and (2) identifying and protecting climate refugia where at least some species might persist in situ. Connectivity has been the most frequently discussed option, dating back to the seminal paper by Peters and Darling (1985). This chapter focuses on the second option—climate refugia—which is receiving increased attention.

Paleoecologists and biogeographers in the nineteenth and twentieth centuries became interested in climate refugia largely to understand and reconcile anomalies between species-specific dispersal rates and documented poleward migration. Species recolonized northern regions from presumed southern refugia much more quickly than their known dispersal abilities might allow, a problem called "Reid's paradox" after the nineteenth-century British botanist Clement

Reid, who noted the inconsistency (Provan and Bennett 2008). The only plausible mechanisms to resolve the paradox were rare, long-distance dispersal events or the existence of cryptic refugia near the glacial boundary; both mechanisms have supportive evidence. The refugia concept was later expanded to include regional and local refuges from heat waves, droughts, and rising sea levels.

Climate change refugia (hereafter climate refugia) have been defined as "areas that are relatively buffered from contemporary climate change over time and where biodiversity can retreat to and persist in" (Keppel et al. 2024). Identifying and protecting climate refugia is crucial in the face of climate change (Noss 2001). Climate refugia exist along a wide range of spatial scales, from macrorefugia on a regional scale to microrefugia on a local scale. Microrefugia are essentially areas out of equilibrium with the regional climate. Both macrorefugia and microrefugia can maintain relatively stable climates and communities for long periods (Dobrowski 2011).

The recognition of refugia in conservation planning is growing, as they may be the only safe places for species to survive a rapidly changing climate, given dispersal limitations. Identification of refugia can help target conservation efforts toward areas that have historically provided stable and favorable climatic conditions and are likely to remain relatively stable into the future (Keppel et al. 2015; Harrison and Noss 2017; Ebersole et al. 2020; Selwood and Zimmer 2020; Rossetto and Kooyman 2021).

Formation of Climate Refugia in Florida

The existence of climate refugia in Florida may be surprising. The ecology and climate of Florida and the Coastal Plain in general are often misunderstood, with many misconceptions regarding past climate velocity and inundation by high sea-level stands, among other myths (Noss et al. 2015).

Climate Velocity and Terrain Misconception: Climate velocity can be calculated as the ratio of climate change over time to the present change in climate over space (Loarie et al. 2009; Sandel et al. 2011). It is often assumed that Florida's low and relatively flat terrain must result in high climate velocity because species are unable to disperse along topographic gradients to seek favorable microclimates. This topographic limitation suggests low climatic resilience. However, this perspective fails to account for the moderating influence of the ocean, which surrounds Florida on three sides and helps to buffer temperature extremes. Regional climate refugia such as Florida are often centers of endemism (see chapter 1) and are found in nonglaciated mountainous regions as well as on islands and in coastal areas where temperature fluctuations are moderated by mild ocean currents. These areas can be expected to remain relatively stable climatically in the future, although the rapid pace of climate change may still lead to high rates of species loss—just not as high as in nonrefugia (Harrison and Noss 2017).

Elevation and Inundation Misconception: Another common myth is that Florida's low elevation has led to total or near-total inundation by the sea during warm periods, such as the Pleistocene interglacials. This obviously would have a significant impact on terrestrial and freshwater biota. Much of Florida was indeed inundated during Pleistocene interglacials, with sea level 6.6–8.3 m higher than today during the most recent interglacial (Muhs et al. 2011). The highest Pleistocene inundation, some 120,000 years ago, was approximately 6–13 m above present mean sea level (Raymo and Mitrovica 2012). During the middle Pliocene, sea level was even higher, about 35 ± 18 m above present (Dowsett and Cronin 1990). To put these numbers into perspective, the highest elevations in Florida are 105 m in the Panhandle and 95 m in the central peninsula; several major ridges, escarpments, and terraces in Florida exceed 40 m. During high sea-level stands, these highlands were archipelagos fostering evolution in isolation of endemic taxa (Sorrie and Weakley 2001; Noss et al. 2015; see chapter 1).

Nevertheless, Florida's modest terrain makes topography less important for providing refugia compared to many other regions. Thus, broad-scale geospatial analyses of climate resilience based largely on topography, such as the climate resiliency models generated at The Nature Conservancy (Anderson et al. 2023), do not apply well to Florida. Instead, other factors contribute to resilience in Florida, including less severe shifts in climate (low climatic exposure) and the ability to stay close to near-optimal climatic conditions (high climate-tracking potential). These factors, along with temperature buffering by the sea, produce relative stability of climate across much of the lower southeastern Coastal Plain (Noss et al. 2015; Michalak et al. 2020).

Exploring Spatial Scales of Climate Change and Refugia in Florida

This chapter explores three distinct spatial scales of climate refugia in Florida. We integrate complementary methodologies that operate at different scales; relying on a single methodology or scale would not produce as robust an evaluation (Michalak et al. 2020; Morelli et al. 2020).

FLORIDA CLIMATE, ENVIRONMENT, AND ECOLOGY

The State of Florida (170,310 km^2 or 65,758 mi^2) is home to 81 natural communities (FNAI 2010). The climate is predominantly humid to subhumid subtropical and tropical, with extensive coastlines on the Gulf of Mexico and Atlantic Ocean. Florida has distinct wet and dry seasons, with approximately 80% of annual rainfall occurring from May to October and the remaining 20% from November to April (Mulholland et al. 1997). In southern and Central Florida, heavy rains are concentrated in the summer, leading to highly seasonal rainfall patterns; in contrast, northern Florida, especially the Panhandle, has a secondary rainy period in winter due to continental cold fronts, resulting in annual rainfall that is more uniform (Chen and Gerber 1990; Noss 2018). Weather patterns are influenced by the Bermuda High, a semipermanent high-pressure system that dominates the weather in the Florida Peninsula (Chen and Gerber 1990). The broad temperature gradient from north to south, especially in winter, provides conditions suitable to species associated with a variety of climates, from temperate to tropical (Mulholland et al. 1997).

EVIDENCE FOR LONG-TERM CLIMATIC STABILITY IN FLORIDA

The North American Coastal Plain, including Florida, is recognized as a global biodiversity hotspot (see chapter 1). Florida is the most biologically rich portion of the Coastal Plain and ranks second only to California in terms of endemic and imperiled species in the US (Gentry 1986; Estill and Cruzan 2001; Sorrie and Weakley 2001; Noss 2012). Recent research into the climatic, geologic, biogeographic, and evolutionary history of the Coastal Plain has provided valuable new insights (Noss et al. 2015). Florida exhibits distinct zones of endemism related to its climatic and evolutionary history (Estill and Cruzan 2001; Knight et al. 2011). Endemics, especially plants, are generally concentrated on ancient ridge systems where allopatric speciation occurred during historic high sea-level stands, when surrounding lowlands were inundated. Some of these endemics are older than often assumed, probably originating in the Miocene or Pliocene (Germain-Aubrey et al. 2014).

The high degree of endemism in Florida is evidence of antiquity and of long-term climatic stability. Fossil pollen and vertebrate fossils suggest that the southern Coastal Plain has been relatively stable climatically for tens of thousands to perhaps millions of years (Noss 2012). Based on a continuous fossil pollen record from lake sediment cores extending back some 61,000 years in Central Florida (Lake Tulane, Highlands County), Grimm et al. (2006) demonstrated predictable cycles of alternating pine and oak dominance corresponding to warm, wet periods (pine) and dry, cool periods (oak) associated with stadial (glacial advance) and interstadial (glacial retreat) phases, respectively. Yes, Florida was warmer when it was cold to the north. Despite these cycles in abundance of dominant plants, the overall vegetation changed little, with no identifiable northern taxa. Grimm et al. (2006) concluded, "The Pleistocene flora of the Lake Wales Ridge is similar to today. Only the quantities vary."

Long-Term Macrorefugia and Ocean Currents

Florida's relative climatic stability is largely due to its geographic position as a peninsula surrounded by relatively mild and stable ocean currents. First, we focus on the Atlantic Warm Pool (AWP), which includes the Gulf of Mexico, Caribbean Sea, and North Atlantic (fig. 5.1). Over time, global temperatures have varied significantly; during the last glacial maximum, global average temperatures were around 21.0°C, while during warmer interglacial periods they reached a maximum of 31°C. However, the core region of the AWP experienced a deglaciation-period temperature increase of only ~2.5°C (Ziegler et al. 2008). Based on marine sea-surface temperature records, comparison of fossil corals, and coral isotope analysis, the temperature in the reef tracts in Southern Florida has remained relatively constant, with only a 2.6°C increase in temperature over the last 4.2 million years (Brachert et al. 2014). The Intertropical Convergence Zone and the Loop Current in the Gulf of Mexico are affected by the size and northward expansion of the AWP, which in turn affects regional temperature and precipitation stability (Donders et al. 2011). The Loop Current, where summer trade winds transport warm surface waters to the Gulf of Mexico, is influenced by local surface winds and water migration from the Caribbean Sea.

Interestingly, sea-surface temperature records from the Gulf of Mexico show little evidence of dramatic historical climate shifts. Among the major climatic events of the Late Pleistocene were the Younger Dryas, an abrupt but brief cooling event circa 12,900 to 11,700 years ago (Rasmussen et al. 2006), and several Hein-

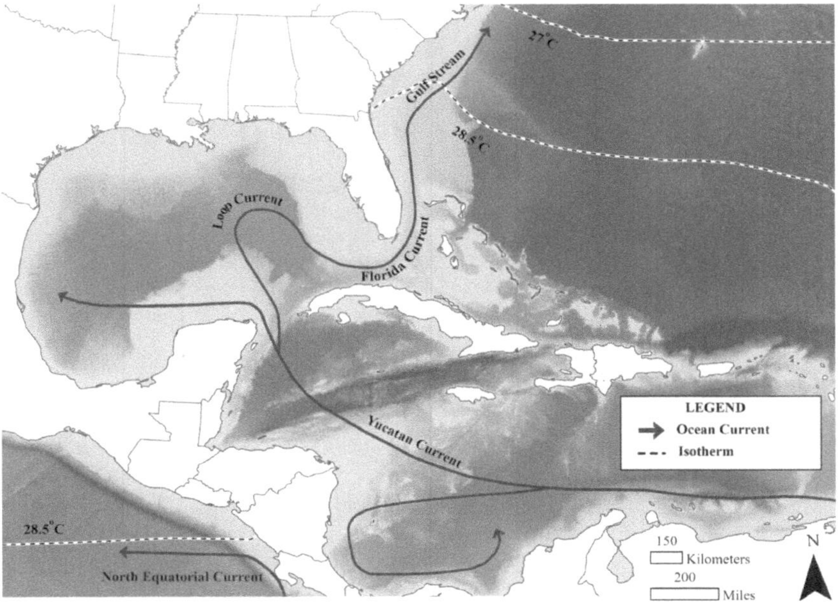

FIGURE 5.1. The Atlantic Warm Pool, indicated by sea-surface temperatures of 28.5°C or higher. Adapted from Donders et al. (2011).

rich Events, characterized by intense iceberg discharge into the North Atlantic (Heinrich 1988). Evidence suggests that the climate in the Gulf region was not greatly affected by these rapid temperature shifts. However, these events influenced precipitation patterns, trade winds, and ocean currents in the subtropical North Atlantic, particularly the Gulf Loop Current (Donders et al. 2011). Florida's climate patterns responded very differently from other areas to these climatic events. During colder stadial periods in the North Atlantic, Florida experienced warmer temperatures, especially in winter (Donders et al. 2011). This surprising pattern can be attributed to the moderating effect of consistently mild ocean currents, which moderated the climate across the state. This is clear evidence that the Florida Peninsula (at least) was a regional climate refugium during the Late Pleistocene.

Lakes in Florida have also been important climate refugia for aquatic taxa and perhaps also for species from surrounding terrestrial communities. Sediment cores indicate that some of Florida's deepest lakes maintained their water levels even during the driest periods of the previous glacial cycle (Hine et al. 2017). For example, Lake Tulane maintained a constant water level over 60,000 years

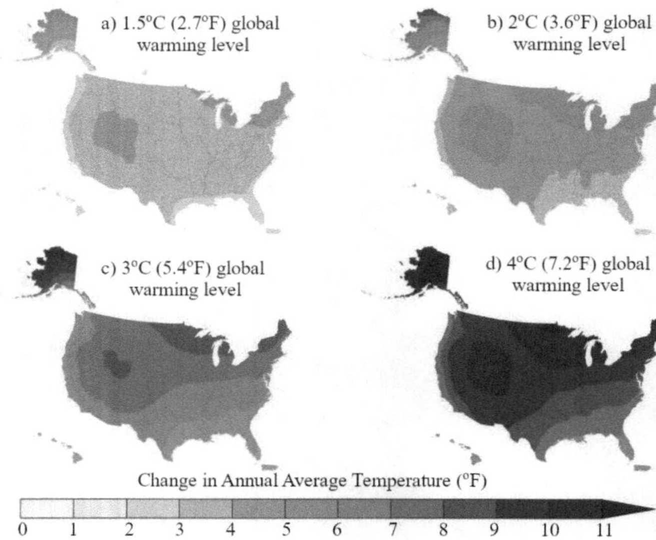

FIGURE 5.2. National climate projections for (*a*) 1.5°C, (*b*) 2.0°C, (*c*) 3.0°C, and (*d*) 4.0°C show Florida is anticipated to undergo the least amount of warming in comparison to the other states. Changes are relative to the period 1851–1900. Based on CMIP6. Adapted with permission from the Fifth Climate Assessment.

(Grimm et al. 2006). From nine to six thousand years ago, increased precipitation contributed to the filling of Florida's numerous shallow lakes, the majority of which have maximum depths of less than 5 m (16 ft) (Hine et al. 2017). Around 11,700 years ago, at the beginning of the Holocene Epoch, the rise in sea level (at a rate up to 4.5 m per century; Donoghue 2011) caused the freshwater karst aquifers beneath Florida's land surface to move upward. This led to the activation of flow in Florida's artesian springs (Hine et al. 2017). This marked a crucial period of environmental change and the beginning of the unique hydrological system in Florida, which we discuss under the section "Identification of Microrefugia."

Identification of Within-State Climate Refugia

According to the Fifth National Climate Assessment, Florida is projected to experience less warming than the rest of the United States (USGCRP 2023) (fig. 5.2). Nevertheless, temperatures will increase. We were interested in determining whether some subregions within Florida are more likely to serve as climate refugia than others. One method for identifying climate refugia on a regional scale is to identify locations with the lowest temperature extremes, as they are likely to remain favorable during periods of extreme heat (Ashcroft et al. 2012). Establishing a baseline is important for measuring environmental stability over time (Radeloff et al. 2015).

We assessed the frequency of extreme summer temperatures and changes in mean summer maximums as proxies for climate change, employing methods adapted from previous studies (Rojas et al. 2021). We applied a dataset with a 250 m spatial resolution from 1950 to 2022, which allowed for detailed examination of climate variability. This PRISM dataset was chosen because of its potential for examination of climate variability at a more detailed scale compared to other data sources with coarser resolutions (Daly et al. 2008). The 250 m data were obtained by downscaling the original dataset with a pixel size of 4 km using the gradient-inverse-distance-squared method (Flint et al. 2013, 2021). These metrics have been extensively utilized in prior evaluations of the impact of climate change on species and ecosystems.

We estimated refugial capacity in Florida by comparing recent temperature data (1980–2022) to a historical baseline (1950–1979) (fig. 1.6). The threshold for extreme summer (June–September) temperature was defined as the 90th percentile of maximal temperatures during the baseline period. A heat wave is defined as a prolonged period during which the highest or lowest temperature exceeds a preset threshold that is at or above the 90th or 99th percentile of the daily temperature distribution (Keellings and Waylen 2014). We used the absolute change in mean summer maximum temperatures as a proxy for climate change, like other studies (Flint et al. 2013; Rojas et al. 2021). The proportion of recent observations exceeding the 90th percentile baseline was calculated for each grid cell; the difference in mean summer temperatures between the two time periods was also determined. The results then were combined into one value and scaled on a gradient from 0 to 1, with 1 indicating areas of the highest potential as climate refugia. Southwest Florida and parts of south-central Florida have shown the greatest increases in mean summer temperatures and frequency of heat waves from 1950–1979 to 1980–2022 (fig. 5.3). Areas with the least change are coastal, again reflecting the moderating influence of the ocean, and mostly in the northern half of the state. Unfortunately, these coastal areas will be most directly impacted by sea-level rise.

Over the past 70 years, Florida has experienced significant increases in summer daily maximum, mean, and minimum heat indexes and wet-bulb globe

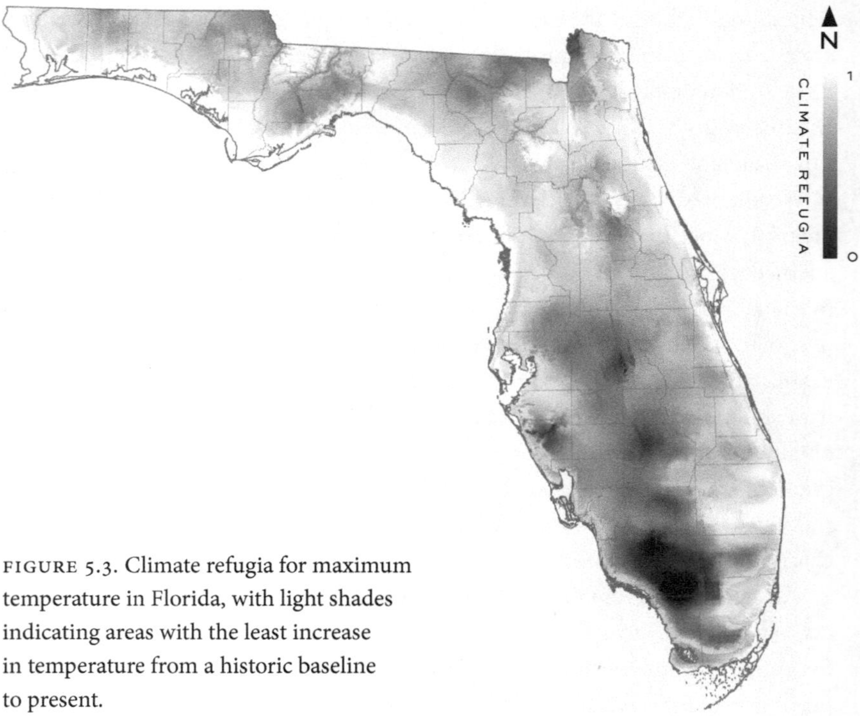

FIGURE 5.3. Climate refugia for maximum temperature in Florida, with light shades indicating areas with the least increase in temperature from a historic baseline to present.

temperatures, with a pronounced rise in nighttime heat stress (McAllister et al. 2022). Furthermore, the incidence of "danger" heat index days (exceeding 104°F or 39.4°C) and "high" wet-bulb globe temperature days (exceeding 88°F or 31.1°C) has markedly increased across the state (McAllister et al. 2022). In certain areas, these extreme heat conditions now constitute over 20% of the summer days (McAllister et al. 2022). Heat wave characteristics exhibit significant spatial variability, but heat waves recently have become more frequent and intense over much of the state (Keellings and Waylen 2014; Cloutier-Bisbee et al. 2019; McAllister et al. 2022). Projection modeling indicates a threefold increase in the frequency and more than a sixfold rise in the average duration of heat waves across Florida, with intensity of heatwaves expected to rise by 4–6°C (Raghavendra et al. 2019). Florida typically experiences monthly or seasonal droughts (Seager et al. 2009). A slight decrease in mean precipitation is projected for the late twenty-first century, alongside an increase in intense heatwave-associated precipitation events, particularly over the central and southern regions of the state (Raghavendra et al. 2019).

Previous analysis showed that temperate species may lose their range in Florida, while some subtropical species may require human intervention to move northward (Crumpacker et al. 2001). Global and regional maps of climate trajectories help predict movements of climate migrants and areas of diminishing species richness (Burrows et al. 2014). More biological models are needed to determine how species' niche requirements connect with microclimates, how far species will need to disperse to suitable microclimates, and what kind of population dynamics might occur in microclimate-suitable habitat areas (Hannah et al. 2014).

Climate change is disrupting ecosystems, leading to negative consequences such as the invasion of nonnative species. In southern Florida, tropical dragonfly species are rapidly invading, suggesting that climate change may cause more range extensions for invasive species (Paulson 2001). Other invasive species with the potential to move northward in Florida include trees (e.g., *Melaleuca quinquenervia*, cajeput), the shrub *Schinus terebinthifolius* (Brazilian pepper), and subtropical/tropical fish species *Clarias batrachus* (walking catfish) (Mulholland et al. 1997). Cyanobacteria, which thrive in temperatures more than 25°C, are causing issues such as eutrophication and loss of water clarity (Mulholland et al. 1997; Paerl and Otten 2013). The water quality in estuaries is influenced by climate; consequently, there are alterations in the transport of nutrients and organic matter as well as modifications in the physical and chemical properties of estuaries (Wetz and Yoskowitz 2013).

Extreme Climate Events in Florida

Climate change is projected to increase the frequency, intensity, and impact of extreme weather events (Maxwell et al. 2019; Neilson et al. 2020), including tropical cyclones, heatwaves, floods, and droughts. A 2°C increase in global temperature could lead to a more than fivefold increase in exposure to these events (Lange et al. 2020). Florida is highly vulnerable to extreme weather events throughout the year but especially during the July–November hurricane season. High sea-surface temperatures are linked to more powerful hurricanes, with a 0.5°C increase in sea-surface temperature associated with a 40% increase in hurricane activity (Saunders and Lea 2008).

From 1950 to 2016, the southwest Florida Gulf Coast saw a significant rise in daily and seasonal rainfall extremes, contributing to a wetter climate (Gitau 2018). These findings highlight a recent trend toward more frequent and severe rainfall events. Increased rainfall can elevate flood risks, soil erosion, and changes in freshwater availability and quality, with significant implications for water management, infrastructure, and conservation.

Extreme events such as hurricanes and flooding have had significant impacts on species in Florida. Hurricanes have caused 88% tree mortality in longleaf pine savannas (Zampieri et al. 2020) and have led to population declines in red-cockaded woodpeckers (*Dryobates borealis*; Cox and McCormick 2016). Habitat models predict a 41% decrease in the woodpecker population due to hurricanes (Lucash et al. 2022). Additionally, extreme events can transport organic debris and suspended solids into estuaries, leading to hypoxia or anoxia (Wetz and Yoskowitz 2013). These disturbances have resulted in changes to productivity, species interactions, and susceptibility to biological invasions. These effects highlight the need for proactive adaptation strategies in resource management.

Studies have examined how animal species respond to environmental change (Urban 2015; Wiens 2016), with special attention to how populations might survive life-threatening weather events (Buchholz et al. 2019). Research on loggerhead sea turtles (*Caretta caretta*) has shown that precipitation and humidity affect hatchling production, with implications for population persistence (Montero et al. 2018). Stone crabs (*Menippe mercenaria*) in the West Florida Shelf are influenced by changes in seawater pH and temperature (Alaerts et al. 2022). Many climate refugia for currently resident species may be ephemeral, as it is likely that global warming will persist, that precipitation patterns will change, and that seasonal and annual extremes will increase (Morelli et al. 2020).

Identification of Microrefugia

Despite its flat appearance, Florida has complex karst topography with sinkholes, caves, and springs (Polk and Brinkmann 2013), which are significant for their biodiversity (FNAI 2010). Changes in the size and relief of the Florida Peninsula over the past ~2.8 million years (the Quaternary Period) have been driven by climatic factors rather than tectonic forces. These climatic drivers have led to a surficial aquifer system as well as fluctuations in the water table and karstification at different elevations within the state (Lace and Mylroie 2013). Florida's rocks are young, eogenetic carbonates (48–25 million years old) with high porosity and permeability (Polk and Brinkmann 2013). Research in other regions has shown that karst dolines, such as sinkholes, have a high abundance of cool-adapted plant species (Bátori et al. 2017).

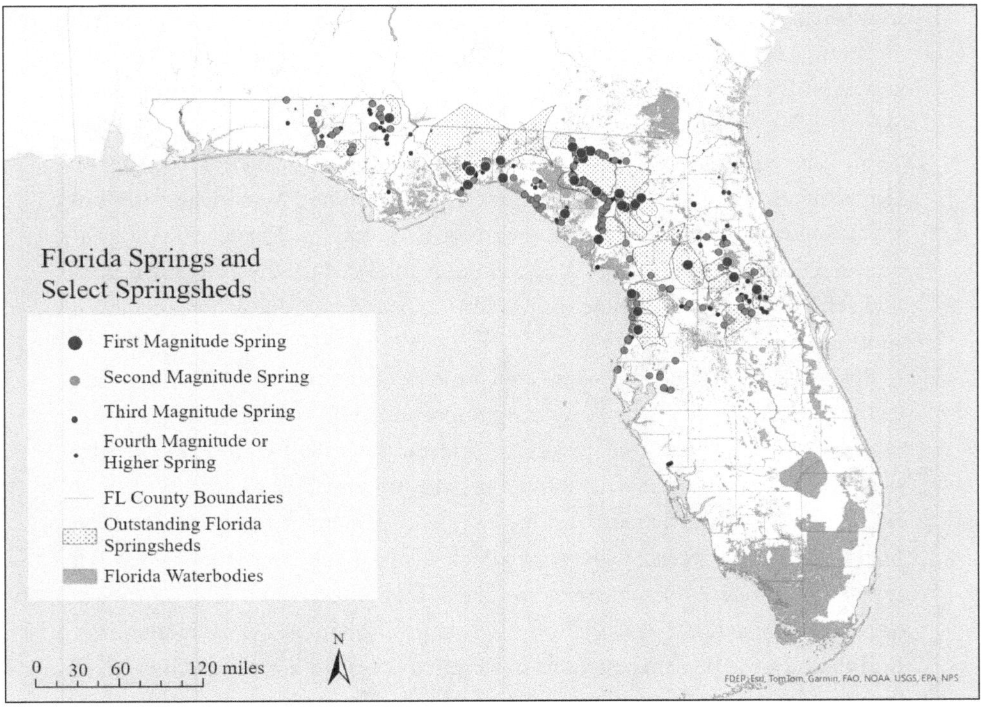

FIGURE 5.4. Florida springs and selected springsheds.

Springs (fig. 5.4) serve as thermal microrefugia because they provide stable and relatively consistent environmental conditions (Cartwright et al. 2020). The potential of springs and spring complexes to serve as stable, relative, or transient refugia depends on specific environmental conditions. Florida has more than 1,000 artesian springs (see chapter 6), with the first-magnitude springs most significant, producing a daily average discharge of over 30 billion liters (8 billion gallons) (Polk and Brinkmann 2013). Additionally, some lakes situated along Florida's high central "spine" reach depths of 20–30 m (67–98 ft) (Hine et al. 2017). Regions with cooler, mesic microenvironments and elevated water availability have the capacity to alleviate the impacts of climate change (McLaughlin et al. 2017).

In a warming climate, increased temperatures will reduce groundwater levels and surface reservoirs due to higher evaporation rates, as shown by hydrologic simulations (Condon et al. 2020). Recent events in Florida, such as lakes drying up for the first time in recorded history and long periods of low lake levels, may be caused by both climate change and human water withdrawals, although substantial evidence suggests that persistent rainfall deficits are usually the main cause (Hine et al. 2017). The impact of even moderate warming on groundwater storage and evaporation is significant and immediate (Condon et al. 2020).

Florida's springs are crucial habitats for numerous local endemic species (Knight et al. 2011), providing stable flow, temperature, and water quality (Odum 1957). These springs maintain a consistent temperature throughout the year between 20.6 and 22.8°C (69–73°F) (Schmidt 2004). Thus, they provide a thermal refuge for cold-intolerant species in winter and heat-intolerant species in summer.

Springs are expected to be one of the most resilient microhabitats in Florida in the face of climate change, offering important refugia for not only aquatic organisms but also terrestrial species in the immediate vicinity. The use of springs as winter refugia for manatees has significantly increased in recent years (Reep and Bonde 2021), particularly due to the closure of power plants that provided warm-water discharges. However, the decline in aquatic vegetation, which is a critical food source for manatees, and high rates of recreation in springs have negatively impacted their suitability as overwinter habitat (Slone et al. 2023). Additionally, the presence of dams has limited access to key refugia like Silver Springs. These factors combined have contributed to massive manatee die-offs in recent years. Spring temperatures adjust to average air temperatures, rising as air temperatures increase. Despite this, springs continue to be important thermal refugia because they discharge cool groundwater into surface water bodies (lakes, rivers, and estuaries) that respond more rapidly to climate fluctuations.

Climate change impacts pose significant threats to springs and their refugia, including increased frequency of severe hurricanes and other major storms. Large rain events can cause springs to cease or reverse flow, leading to stagnant pools or sinkholes (Brown et al. 2014). Each spring has a varied level of vulnerability to flood disturbances based on its morphological and hydrological characteristics (Donsky et al. 2023), making it crucial to assess their resilience as climate refugia. Flood disturbances occur along a gradient of magnitude and duration ranging from several days to months (Hensley and Cohen 2017).

Sea-Level Rise

Sea levels are rising due to global warming and the melting of ice sheets (IPCC 2013). The global mean sea level has increased by 16 to 21 cm in the past century

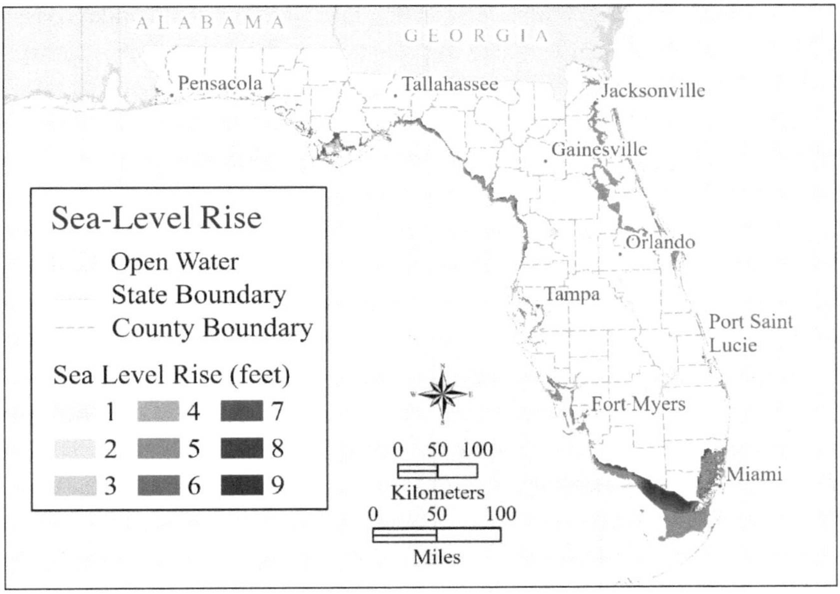

FIGURE 5.5. Sea-level rise projections for Florida. Data source: 2009 digital elevation model by the Florida Department of Environmental Protection.

and is expected to rise even more and faster in the future. The IPCC projects that global mean sea level will rise up to 1 m by 2100 (Vaidya et al. 2023), although other estimates indicate a rise of 2 m or more by 2100 (Oppenheimer and Alley 2016). Sea-level rise (SLR) poses significant risks to coastal regions, particularly in Florida, where over half of the population lives along its extensive 1,931 km (1,200 mi) coastline (Milordis et al. 2023).

SLR is impacting not only cities but also ecosystems. It is causing more frequent coastal flooding, saltwater intrusion into freshwater systems, stronger storm surges, and decline of coastal ecosystems like mangroves and salt marshes. Additionally, SLR is leading to the recession of beaches and dunes (Walton 2007; Milordis et al. 2023). Coastal ecosystems, the Everglades, and the Florida Keys are expected to undergo significant transformations due to rising sea levels (Noss 2011; Wells et al. 2014) (fig. 5.5).

Florida's coastal zones are at risk of flooding due to their low elevation and susceptibility to heavy rainfall, high tides, and storm surges. Barrier islands and exposed coastline are particularly vulnerable to erosion, overwashing, and breaches. The geological composition of loose sand and porous limestone worsens the effects of saltwater intrusion and flooding. About 50% of Florida's coastal zone is now urbanized, increasing the vulnerability to SLR and other climate-related stressors (Parkinson and Wdowinski 2023).

Model forecasts predict that the Gulf of Mexico's average sea-surface temperature will increase by 1.94°C between 2001 and 2100, while the Atlantic Ocean's average temperature will rise by 0.55°C. The mean sea level is projected to rise by 51 cm in the Gulf and 48 cm in the Atlantic (Chang et al. 2014). This increase is expected to result in approximately 7,770 km^2 (3,000 mi^2) of flooded land in Florida by 2100, which is equivalent to 5.2% of the state's current land area. Most of the flooding occurs in the southern region, with the Everglades experiencing the most significant impact (Chang et al. 2014).

Parkinson and Wdowinski (2023) analyzed sea-level data from 14 NOAA tide gauge stations in Florida and found the average rate of SLR has been increasing over time. Historical rates ranged from 2.2 to 4.2 mm per year, averaging 3.1 mm. Since 1993, rates ranged from 4.7 to 6.9 mm per year, averaging 5.9 mm. From 2003 to 2022, rates ranged from 7.6 to 10.0 mm per year, averaging 8.2 mm. The study suggests SLR will continue to accelerate during this century, leading to increased inland movement or submersion of Florida's coastal environments (Parkinson and Wdowinski 2023). The findings highlight the need for proactive measures to mitigate the impacts of SLR, including resilient infrastructure and coastal habitat restoration. Continued research and monitoring are also crucial for informing adaptation strategies.

Conservation planning can involve creating corridors of suitable habitat and implementing infrastructure modifications to help species and communities affected by 1+ m SLR. These corridors can serve as pathways for species to escape rising seas (Noss 2011). Alternative futures in Florida have been projected using various SLR scenarios to illustrate the consequences on species. Sea-level rise is causing shorebird populations to collapse due to nest flooding long before their habitats drown (van de Pol et al. 2024). Projections show significant Florida panther (*Puma concolor*) and black bear (*Ursus americanus floridanus*) habitat submerged by 2070 (Davis et al. 2021). The rise in sea level is also expected to affect nesting habitat for American crocodiles (*Crocodylus acutus*) in Florida by inundating a significant portion of their habitat.

Saltwater intrusion into the aquifer is a significant threat in Florida, as no location is farther than 100 km from the coast (Romañach et al. 2020). The coastal karst aquifer is highly susceptible to seawater contamination through subsurface conduit networks, with the Woodville Karst Plain in Florida experiencing the longest distance intrusion of seawater (Xu et al. 2016). The Biscayne Aquifer, which includes a substantial portion of the Everglades and the Miami metropolitan area, is submerged due to a canal network that lowered the water table, allowing saltwater to enter the interior (Lace and Mylroie 2013).

Conclusions and Critical Conservation Actions Needed

Climate refugia in Florida might provide safe havens for species to survive and adapt to climate change. By focusing on refugia, conservation efforts can be directed toward areas that are highly likely to maintain relatively stable and favorable climate conditions for species into the future (Keppel et al. 2015, 2024; Ebersole et al. 2020; Selwood and Zimmer 2020; Rossetto and Kooyman 2021). Historical climate patterns and local factors such as hydrology and microtopography have created diverse types of refugia, ranging from macrorefugia at a subcontinental scale to ridges at a regional scale to microrefugia at a local scale. Identifying and protecting refugia is a critical part of landscape conservation planning. It requires assessment of factors such as climate stability, ecosystem resilience, and the severity of future climate-related and land-use threats. We have identified several regions and types of climate refugia within Florida, but much more field inventory, niche modeling, mapping, and other research are needed to locate those subregions and local sites with the greatest refugial potential.

Florida's climate refugia are influenced by ocean currents, historical climatic events, and the unique karst geology of the state. However, climate change, including sea-level rise, increased frequency of extreme weather events, and saline intrusion, still poses significant threats to refugia. Landscape conservation planning for biodiversity should prioritize the identification, protection, and enhancement of climate refugia, along with strategies such as landscape connectivity to facilitate adaptive shifts in species ranges.

We recognize that protection of climate refugia may seem at odds with implementation of statewide ecological networks, such as the Florida Wildlife Corridor, which are based on the needs of wide-ranging focal species and on expansive ecosystem services such as watershed protection. Many of the most important refugia, such as springs, caves, sinkholes, and steep ravines, are localized and isolated, and cannot easily be accommodated within regional networks of conservation areas and broad corridors. Ridges, which serve as refugia from extreme sea-level rise, are mostly heavily developed in Florida, with only small fragments of natural habitat remaining (the Lake Wales Ridge and Miami Rock

Ridge are prime examples). These refugia require a complementary strategy of small-site protection—which, as described in chapter 1, is key to protecting many of Florida's rarest endemic species. A comprehensive regional landscape approach to conservation must strive to protect both large core areas and connecting corridors as well as small, isolated sites of high conservation value. This combined strategy is essential for both biodiversity protection and adaptation to climate change.

Literature Cited

Alaerts, L., T. Dobbelaere, P. M. Gravinese, and E. Hanert. 2022. Climate change will fragment Florida stone crab communities. Frontiers in Marine Science 9:839767.

Anderson, M. G., M. Clark, A. P. Olivero, A. R. Barnett, K. R. Hall, M. W. Cornett, et al. 2023. A resilient and connected network of sites to sustain biodiversity under a changing climate. Proceedings of the National Academy of Sciences 120:e2204434119.

Ashcroft, M. B., J. R. Gollan, D. I. Warton, and D. Ramp. 2012. A novel approach to quantify and locate potential microrefugia using topoclimate, climate stability, and isolation from the matrix. Global Change Biology 18:1866–1879.

Bátori, Z., A. Vojtkó, T. Farkas, A. Szabó, K. Havadtői, A. E. Vojtkó, et al. 2017. Large- and small-scale environmental factors drive distributions of cool-adapted plants in karstic microrefugia. Annals of Botany 119:301–309.

Brachert, T. C., M. Reuter, S. Krüger, H. Lohmann, E. J. Petuch, and J. S. Klaus. 2014. A 4.2 million year record of interglacial paleoclimate from sclerochronological data of Florida carbonate platform (Early Pliocene to recent). Global and Planetary Change 12054–12064.

Brown, A. L., J. B. Martin, E. J. Screaton, J. E. Ezell, P. Spellman, and J. Gulley. 2014. Bank storage in karst aquifers: The impact of temporary intrusion of river water on carbonate dissolution and trace metal mobility. Chemical Geology 385:56–69.

Buchholz, R., J. D. Banusiewicz, S. Burgess, S. Crocker-Buta, L. Eveland, and L. Fuller. 2019. Behavioural research priorities for the study of animal response to climate change. Animal Behaviour 150:127–137.

Burrows, M. T., D. S. Schoeman, A. J. Richardson, J. G. Molinos, A. Hoffmann, L. B. Buckley, et al. 2014. Geographical limits to species-range shifts are suggested by climate velocity. Nature 507:492–495.

Cartwright, J. M., K. A. Dwire, Z. Freed, S. J. Hammer, B. McLaughlin, L. W. Misztal, et al. 2020. Oases of the future? Springs as potential hydrologic refugia in drying climates. Frontiers in Ecology and the Environment 18:245–253.

Chang, B., J. Guan, and M. M. Aral. 2014. Modeling spatial variations of sea level rise and corresponding inundation impacts: A case study for Florida, USA. Water Quality, Exposure and Health 6:39–51.

Chen, E., and J. F. Gerber. 1990. Climate. Pages 11–34 in R. L. Myers and J. J. Ewel, editors. Ecosystems of Florida. University of Central Florida Press, Orlando.

Cloutier-Bisbee, S. R., A. Raghavendra, and S. M. Milrad. 2019. Heat waves in Florida: Climatology, trends, and related precipitation events. Journal of Applied Meteorology and Climatology 58:447–466.

Condon, L. E., A. L. Atchley, and R. M. Maxwell. 2020. Evapotranspiration depletes groundwater under warming over the contiguous United States. Nature Communications 11:873.

Cox, J. A., and J. K. McCormick. 2016. New insights from an attempt to reintroduce Red-cockaded Woodpeckers in northern Florida. Journal of Field Ornithology 87:360–370.

Crumpacker, D. W., E. O. Box, and E. D. Hardin. 2001. Implications of climatic warming for conservation of native trees and shrubs in Florida. Conservation Biology 15:1008–1020.

Daly, C., M. Halbleib, J. I. Smith, W. P. Gibson, M. K. Doggett, G. H. Taylor, et al. 2008. Physiographically sensitive mapping of climatological temperature and precipitation across the conterminous United States. International Journal of Climatology 28:2031–2064.

Davis, A. G., J. J. Cox, and S. Fei. 2021. Alternative 2070: Mitigating the effects of projected sea level rise and urbanization on Florida black bear and Florida panther habitat. Journal for Nature Conservation 63:126052.

Dobrowski, S. Z. 2011. A climatic basis for microrefugia: The influence of terrain on climate. Global Change Biology 17:1022–1035.

Donders, T. H., H. J. De Boer, W. Finsinger, E. C. Grimm, S. C. Dekker, G. J. Reichart, et al. 2011. Impact of the Atlantic Warm Pool on precipitation and temperature in Florida during North Atlantic cold spells. Climate Dynamics 36:109–118.

Donoghue, J. F. 2011. Sea level history of the northern Gulf of Mexico coast and sea level rise scenarios for the near future. Climatic Change 107:17–33.

Donsky, P., M. J. Cohen, S. Howley, and G. Klarenberg. 2023. Flood disturbances impact the autotrophic communities in the karst springs of the Suwannee River, FL. Presented at the Water Institute Symposium. University of Florida, Gainesville.

Dowsett, H. J., and T. M. Cronin. 1990. High eustatic sea level during the middle Pliocene: Evidence from the southeastern U.S. Atlantic Coastal Plain. Geology 18:435–438.

Ebersole, J. L., R. M. Quiñones, S. Clements, and B. H. Letcher. 2020. Managing climate refugia for freshwater fishes under an expanding human footprint. Frontiers in Ecology and the Environment 18:271–280.

Estill, J., and M. Cruzan. 2001. Phytogeography of rare plant species endemic to the Southeastern United States. Castanea 66:3–23.

Flint, L. E., A. L. Flint, and M. Stern. 2021. The Basin Characterization Model—A Regional Water Balance Software Package. https://doi.org/10.3133/tm6H1.

Flint, L. E., A. L. Flint, J. H. Thorne, and R. Boynton. 2013. Fine-scale hydrologic modeling for regional landscape applications: The California Basin Characterization Model development and performance. Ecological Processes 2:25.

Florida Natural Areas Inventory (FNAI). 2010. Guide to the Natural Communities of Florida. https://www.fnai.org/species-communities/natcom-guide. Accessed August 7, 2025.

Gentry, A. H. 1986. Endemism in tropical versus temperate plant communities. Pages 153–181 in M. E. Soulé, editor. Conservation Biology: The Science of Scarcity and Diversity. Sinauer, Sunderland, Massachusetts.

Germain-Aubrey, C. C., P. S. Soltis, K. M. Neubig, T. Thurston, D. E. Soltis, and M. A. Gitzendanner. 2014. Using comparative biogeography to retrace the origins of an ecosystem: The case of four plants endemic to the Central Florida scrub. International Journal of Plant Sciences 175:418–431.

Gitau, M. 2018. Patterns in indices of daily and seasonal rainfall extremes: Southwest Florida Gulf Coastal Zone. Climate 6:83.

Grimm, E. C., W. A. Watts, G. L. Jacobson Jr., B. C. S. Hansen, H. R. Almquist, and A. C. Dieffenbacher-Krall. 2006. Evidence for warm wet Heinrich events in Florida. Quaternary Science Reviews 25:2197–2211.

Hannah, L., L. Flint, A. D. Syphard, M. A. Moritz, L. B. Buckley, and I. M. McCullough. 2014. Fine-grain modeling of species' response to climate change: Holdouts, stepping-stones, and microrefugia. Trends in Ecology and Evolution 29:390–397.

Harrison, S., and R. Noss. 2017. Endemism hotspots are linked to stable climatic refugia. Annals of Botany 119:207–214.

Heinrich, H. 1988. Origin and consequences of cyclic ice rafting in the northeast Atlantic Ocean during the past 130,000 years. Quaternary Research 29:142–152.

Hensley, R. T., and M. J. Cohen. 2017. Flow reversals as a driver of ecosystem transition in Florida's springs. Freshwater Science 36:14–25.

Hine, A., E. Martin, J. Jaeger, and M. Brenner. 2017. Paleoclimate of Florida. Pages 105–138 in E. P. Chassignet, J. W. Jones, V. Mirsa, and J. Obeysekera, editors. Florida's Climate. Florida Climate Institute, Gainesville.

IPCC. 2013. Summary for Policymakers. Climate Change 2013: The Physical Science Basis. Contribution of Working Group I to the Fifth Assessment Report of the Intergovernmental Panel on Climate Change. Cambridge University Press, Cambridge, UK, and New York, USA.

Keellings, D., and P. Waylen. 2014. Increased risk of heat waves in Florida: Charac-

terizing changes in bivariate heat wave risk using extreme value analysis. Applied Geography 46:90–97.
Keppel, G., K. Mokany, G. W. Wardell-Johnson, B. L. Phillips, J. A. Welbergen, and A. E. Reside. 2015. The capacity of refugia for conservation planning under climate change. Frontiers in Ecology and the Environment 13:106–112.
Keppel, G., D. Stralberg, T. L. Morelli, and Z. Bátori. 2024. Managing climate-change refugia to prevent extinctions. Trends in Ecology and Evolution 39:800–808.
Keppel, G., K. P. Van Niel, G. W. Wardell-Johnson, C. J. Yates, M. Byrne, L. Mucina, et al. 2012. Refugia: Identifying and understanding safe havens for biodiversity under climate. Global Ecology and Biogeography 21:393–404.
Kim, D., S.-K. Lee, H. Lopez, G. R. Foltz, C. Wen, R. West, et al. 2023. Increase in Cape Verde hurricanes during Atlantic Niño. Nature Communications 14:3704.
Knight, G. R., J. Oetting, and L. Cross. 2011. Atlas of Florida's Natural Heritage: Biodiversity, Landscapes, Stewardship, and Opportunities. Institute of Science and Public Affairs, Florida State University, Tallahassee.
Lace, M. J., and J. E. Mylroie, editors. 2013. Coastal Karst Landforms. https://link.springer.com/10.1007/978-94-007-5016-6. Springer Netherlands, Dordrecht. Accessed January 9, 2024.
Lange, S., J. Volkholz, T. Geiger, F. Zhao, I. Vega, and T. Veldkamp. 2020. Projecting exposure to extreme climate impact events across six event categories and three spatial scales. Earth's Future 8:e2020EF001616.
Loarie, S. R., P. B. Duffy, H. Hamilton, G. P. Asner, C. B. Field, and D. D. Ackerly. 2009. The velocity of climate change. Nature 462:1052–1055.
Lucash, M. S., S. Weiss, M. J. Duveneck, and R. M. Scheller. 2022. Managing for red-cockaded woodpeckers is more complicated under climate change. The Journal of Wildlife Management 86:e22309.
Maxwell, S. L., N. Butt, M. Maron, C. A. McAlpine, S. Chapman, A. Ullmann, et al. 2019. Conservation implications of ecological responses to extreme weather and climate events. Diversity and Distributions 25:613–625.
McAllister, C., A. Stephens, and S. M. Milrad. 2022. The heat is on: Observations and trends of heat stress metrics during Florida summers. Journal of Applied Meteorology and Climatology 61:277–296.
McLaughlin, B. C., D. D. Ackerly, P. Z. Klos, J. Natali, T. E. Dawson, and S. E. Thompson. 2017. Hydrologic refugia, plants, and climate change. Global Change Biology 23:2941–2961.
Michalak, J. L., D. Stralberg, J. M. Cartwright, and J. J. Lawler. 2020. Combining physical and species-based approaches improves refugia identification. Frontiers in Ecology and the Environment 18:254–260.
Milordis, A. C., W. H. Butler, and T. J. Holmes. 2023. What is slowing progress on cli-

mate change adaptation? Evaluating barriers to planning for sea level rise in Florida. Mitigation and Adaptation Strategies for Global Change 28:42.

Montero, N., S. A. Ceriani, K. Graham, and M. M. P. B. Fuentes. 2018. Influences of the local climate on loggerhead hatchling production in North Florida: Implications from climate change. Frontiers in Marine Science 5:262.

Morelli, T. L., C. W. Barrows, A. R. Ramirez, J. M. Cartwright, D. D. Ackerly, T. D. Eaves, et al. 2020. Climate-change refugia: Biodiversity in the slow lane. Frontiers in Ecology and the Environment 18:228–234.

Muhs, D. R., K. R. Simmons, R. R. Schumann, and R. B. Halley. 2011. Sea-level history of the past two interglacial periods: New evidence from U-series dating of reef corals from south Florida. Quaternary Science Reviews 30:570–590.

Mulholland, P. J., G. R. Best, C. C. Coutant, G. M. Hornberger, J. L. Meyer, P. J. Robinson, et al. 1997. Effects of climate change on freshwater ecosystems of the south-eastern United States and the Gulf Coast of Mexico. Hydrological Processes 11:949–970.

Neilson, E. W., C. T. Lamb, S. M. Konkolics, M. J. L. Peers, Y. N. Majchrzak, D. Doran-Myers, et al. 2020. There's a storm a-coming: Ecological resilience and resistance to extreme weather events. Ecology and Evolution 10:12147–12156.

Noss, R. F. 2001. Beyond Kyoto: Forest management in a time of rapid climate change. Conservation Biology 15:578–590.

Noss, R. F. 2011. Between the devil and the deep blue sea: Florida's unenviable position with respect to sea level rise. Climatic Change 107:1–16.

Noss, R. F. 2012. Forgotten Grasslands of the South: Natural History and Conservation. Island Press, Washington, DC.

Noss, R. F. 2018. Fire Ecology of Florida and the Southeastern Coastal Plain. University Press of Florida, Gainesville.

Noss, R. F., W. J. Platt, B. A. Sorrie, A. S. Weakley, D. B. Means, J. Costanza, and R. K. Peet. 2015. How global biodiversity hotspots may go unrecognized: Lessons from the North American Coastal Plain. Diversity and Distributions 21:236–244. https://doi.org/10.1111/ddi.12278.

Odum, H. T. 1957. Trophic structure and productivity of Silver Springs, Florida. Ecological Monographs 27:55–112.

Oppenheimer, M., and R. B. Alley. 2016. How high will the seas rise? Science 354:1375–1377.

Paerl, H. W., and T. G. Otten. 2013. Harmful cyanobacterial blooms: Causes, consequences, and controls. Microbial Ecology 65:995–1010.

Parkinson, R. W., and S. Wdowinski. 2023. A unified conceptual model of coastal response to accelerating sea level rise, Florida, U.S.A. Science of the Total Environment 892:164448.

Paulson, D. R. 2001. Recent Odonata records from southern Florida: Effects of global warming? International Journal of Odonatology 4:57–69.

Peters, R. L., and J. D. S. Darling. 1985. The greenhouse-effect and nature reserves. BioScience 35:707–717.
Polk, J., and R. Brinkmann. 2013. Climatic influences on coastal cave and karst development in Florida. Coastal Karst Landforms: 429. Springer Science, Dordrecht.
Provan, J., and K. D. Bennett. 2008. Phylogeographic insights into cryptic glacial refugia. Trends in Ecology and Evolution 23:564–571.
Radeloff, V. C., J. W. Williams, B. L. Bateman, K. D. Burke, S. K. Carter, E. S. Childress, et al. 2015. The rise of novelty in ecosystems. Ecological Applications 25:2051–2068.
Raghavendra, A., A. Dai, S. M. Milrad, and S. R. Cloutier-Bisbee. 2019. Floridian heatwaves and extreme precipitation: Future climate projections. Climate Dynamics 52:495–508.
Rasmussen, S. O., K. K. Andersen, A. M. Svensson, J. P. Steffensen, B. M. Vinther, H. B. Clausen, M.-L. Siggaard-Andersen, S. J. Johnsen, L. B. Larsen, D. Dahl-Jensen, and M. Bigler. 2006. A new Greenland ice core chronology for the last glacial termination. Journal of Geophysical Research 111. https://doi.org/10.1029/2005JD006079.
Raymo, M. E., and J. X. Mitrovica. 2012. Collapse of polar ice sheets during the stage 11 interglacial. Nature 483:453–456.
Reep, R., and R. Bonde. 2021. The Florida Manatee: Biology and Conservation. Second Edition. University Press of Florida, Gainesville.
Rojas, I. M., M. K. Jennings, E. Conlisk, A. D. Syphard, J. Mikesell, A. M. Kinoshita, et al. 2021. A landscape-scale framework to identify refugia from multiple stressors. Conservation Biology 36:e13834.
Romañach, S. S., A. M. Benscoter, and S. M. Haider. 2020. Potential impacts of future urbanization and sea level rise on Florida's natural resources. Journal of Fish and Wildlife Management 11:174–184.
Rossetto, M., and R. Kooyman. 2021. Conserving refugia: What are we protecting and why? Diversity 13:67.
Sandel, B., L. Arge, B. Dalsgaard, R. G. Davies, K. J. Gaston, W. J. Sutherland, et al. 2011. The influence of Late Quaternary climate-change velocity on species endemism. Science 334:660–664.
Saunders, M. A., and A. S. Lea. 2008. Large contribution of sea surface warming to recent increase in Atlantic hurricane activity. Nature 451:557–560.
Schmidt, W. 2004. Springs of Florida. Bulletin No 66. Florida Geological Survey, Tallahassee.
Seager, R., A. Tzanova, and J. Nakamura. 2009. Drought in the southeastern United States: Causes, variability over the last millennium, and the potential for future hydroclimate change. Journal of Climate 22:5021–5045.
Selwood, K. E., and H. C. Zimmer. 2020. Refuges for biodiversity conservation: A review of the evidence. Biological Conservation 245:108502.
Slone, D. H., S. M. Butler, J. P. Reid, J. Kleen, and J. Palmer. 2023. How do ambient

conditions and management actions affect manatee movements and habitat use? Journal of Wildlife Management 87:e22411.
Sorrie, B. A., and A. S. Weakley. 2001. Coastal Plain plant endemics: Phytogeographic patterns. Castanea 66:50–82.
Urban, M. C. 2015. Accelerating extinction risk from climate change. Science 348:571–573.
USGCRP. 2023. Fifth National Climate Assessment. A. R. Crimmins, C. W. Avery, D. R. Easterling, K. E. Kunkel, B. C. Stewardt, and T. K. Maycock, editors. US Global Change Research Program, Washington, DC. https://repository.library.noaa.gov/view/noaa/61592. Accessed August 7, 2025.
Vaidya, H. N., R. D. Breininger, M. Madrid, S. Lazarus, and N. N. Kachouie. 2023. Generalized additive models for predicting sea level rise in coastal Florida. Geosciences 13:310.
van de Pol, M., L. D. Bailey, M. Frauendorf, A. M. Allen, M. van der Sluijs, N. Hijner, L. Brouwer, H. de Kroon, E. Jongejans, and B. J. Ens. 2024. Sea-level rise causes shorebird population collapse before habitats drown. Nature Climate Change 14:839–844.
Walton, T. L. 2007. Projected sea level rise in Florida. Ocean Engineering 34:1832–1840.
Wells, O., W. Gray, and P. Monaghan. 2014. Adaptation approaches to sea-level rise in Florida: EDIS 2014(7).
Wetz, M. S., and D. W. Yoskowitz. 2013. An 'extreme' future for estuaries? Effects of extreme climatic events on estuarine water quality and ecology. Marine Pollution Bulletin 69:7–18.
Wiens, J. J. 2016. Climate-related local extinctions are already widespread among plant and animal species. PLOS Biology 14:e2001104.
Xu, Z., S. W. Bassett, B. Hu, and S. B. Dyer. 2016. Long distance seawater intrusion through a karst conduit network in the Woodville Karst Plain, Florida. Scientific Reports 6:32235.
Zampieri, N. E., S. Pau, and D. K. Okamoto. 2020. The impact of Hurricane Michael on longleaf pine habitats in Florida. Scientific Reports 10:8483.
Ziegler, M., D. Nürnberg, C. Karas, R. Tiedemann, and L. J. Lourens. 2008. Persistent summer expansion of the Atlantic Warm Pool during glacial abrupt cold events. Nature Geoscience 1:601–605.

6

Applying a Landscape Approach to Aquatic and Coastal Conservation

Aaron Adams, JoEllen Wilson, Philip Stevens, and Emily Taylor

Why a Landscape Conservation Approach Is Needed for Coastal Fisheries

Florida's coastal habitats are rapidly declining due to coastal and upland development. Anthropogenic activities are leading to increased pollution, modified watershed hydrology, and habitat alteration (loss, degradation, and fragmentation). As Florida's surface and groundwaters move from inland to the coasts, they flow through increasingly fragmented landscapes bisected by agriculture, residential developments, and urbanized areas. Extreme hydrologic modifications have occurred in the southern half of the state, most notably at Lake Okeechobee and the Everglades. Historically, water from the lake flowed south in a sheet flow pattern through the Everglades into Florida Bay. A lock and dam system was built in the 1930s to drain water east through the Saint Lucie River and west through the Caloosahatchee River. As a result, Florida Bay now receives considerably less freshwater, resulting in record hypersalinity events exceeding 66 ppt and causing large-scale seagrass loss (Fredley et al. 2019).

High freshwater inflow events associated with Lake Okeechobee discharges have altered environmental conditions on the east coast, which has disrupted seasonal signatures and decoupled forage fish guilds and sportfish (Lewis and

Cook 2023). On the west coast, altered freshwater flows have led to homogenization of the fauna where species most resilient to altered freshwater inputs are favored instead of historically expected seasonal changes in fish assemblages (Olin et al. 2015). Additionally, discharges rich in anthropogenic nutrients from Lake Okeechobee are causing blue-green algae blooms that flow into estuaries on both coasts. This is exacerbating red tides on the west coast (Medina et al. 2022; Phlips et al. 2023), resulting in extensive fish kills and fishery closures by the Florida Fish and Wildlife Conservation Commission (FWC 2024a) and decreased tourism for local beach communities (Heil et al. 2021).

Florida's water and habitat problems are causing coastal systems to reach or exceed tipping points. For example, nutrient loads that reach a threshold where environmental conditions change suddenly cause ecological shifts that may be difficult to reverse (Caddy 2000; Hagy et al. 2004; Duarte et al. 2009). Excessive algal blooms with cascading effects (e.g., seagrass losses from shading by algae, fish kills resulting from low dissolved oxygen) have occurred in numerous estuaries in Florida, including the northern portions of the Indian River Lagoon (IRL) along the east coast. Seagrass cover (percent) in the IRL began to decline in the early 2000s due to eutrophication and exceeded a threshold with massive algal blooms beginning in 2010 (Morris et al. 2022). Seagrass percent cover was >30% in 2000, declined to 17% in 2010, and is now <5% after a decade of intense, frequent, and long-duration harmful algal blooms (HABs).

This loss of essential fish habitat can affect the resilience and rate of recovery of fisheries. For example, the common snook (*Centropomus undecimalis*) population in the northern IRL took more than 10 years to recover from mortality caused by a severe cold event in 2010, while snook populations in other estuaries recovered within four years (Boucek et al. 2023). Emerging issues that also contribute to water quality–associated threats include per- and polyfluoroalkyl substances (PFAS), microplastics, and pharmaceuticals. For example, muscle tissue in harvested fish species in an estuary in southwest Florida exceeded consumption thresholds for PFAS (Pulster et al. 2022), while 50% of juvenile snook in a Brazilian estuary were found to have ingested microplastics (Ferreira et al. 2019). Bonefish (*Albula vulpes*) in South Florida contain on average 7.1 pharmaceuticals, and 50% of fish sampled had at least one pharmaceutical in high enough quantities for a therapeutic effect (Castillo et al. 2023). Red drum (*Sciaenops ocellatus*) sampled from nine estuaries statewide contained an average of 2–4 pharmaceuticals per fish.

At the seascape scale, habitat alteration and altered hydrology of rivers and tidal creeks can result in habitat fragmentation. Over 35% of mangroves have been destroyed worldwide (Polidoro et al. 2010), and an estimated 50% of estuaries have been altered to some extent by humans (Jung et al. 2024). These alterations directly impacted productivity of fish populations and hindered recovery of populations that have declined (Stamp et al. 2022). This is because many fishes exhibit ontogenetic habitat shifts that require an intact coastal habitat mosaic to support their respective life stages. A habitat mosaic can be defined as an array of multiple, contiguous habitat types—such as mangroves, seagrass, benthic algae, sand, coral rubble, limestone, and mud bottom—that supports an inherently diverse community that includes resident species and species that use these habitats as nurseries (Barbier et al. 2011). For most species, the nursery habitat is most vulnerable because (1) a fish population is dependent on juvenile survival for recruitment, (2) nursery habitats often occur on a small spatial scale, (3) nursery habitats are often close to nearshore human activities such as urban development, and (4) nursery habitats often traverse the margins between uplands and estuaries, so take the brunt of watershed alteration.

Urbanization of tidal creek watersheds results in a flashier hydrograph, elevated concentrations of nutrients and contaminants, altered morphology, reduced biotic richness, and increased dominance of tolerant species (Walsh et al. 2005; O'Leary et al. 2021). Species that use tidal creeks and associated coastal ponds include economically important juvenile snook and tarpon (*Megalops atlanticus*; Wilson et al. 2019, 2023). Adams et al. (2009) found lower prey species diversity in tidal creeks used by juvenile snook that experienced higher alteration of freshwater flows, which negatively impacted snook diet. Moreover, the movement of snook and tarpon from isolated nursery habitats toward the open estuary during ontogenetic habitat shifts is dependent on seasonal changes in hydrology and tropical storm flooding that connect multiple habitat types (Barbour et al. 2014). These connections can be disrupted by habitat and water flow alterations. Conservation of a connected mosaic of habitats that encompasses watersheds and coastal areas is needed to sustain estuarine fisheries (Yee et al. 2015; Woo et al. 2019).

Marine fisheries are managed through stock assessments within geographic jurisdictional boundaries, with stock status affected primarily by estimates of harvest and fishing effort. The importance of including habitat, water quality, and other environmental metrics in marine fisheries management has long been acknowledged, but it has been challenging to incorporate this information into fisheries management. In Florida, strategies for fisheries management of inshore sportfish are evolving to account for the effects of environmental conditions on fish populations. For example, in recent years, emergency management measures were requested by stakeholders—and supported by fish monitoring efforts—to aid in fish population recovery. During 2010–2020, the snook fishery in southwest Florida was completely closed to harvest in 6 of 10 years to enable population recovery following a severe cold event and red tides (FWC 2024a); the red drum fishery in the IRL was closed to harvest due to low population size, the result of extensive seagrass loss and poor water quality (FWC 2024b).

To better address the environmental influences on fish populations, the Division of Marine Fisheries Management of FWC recently implemented a regional management approach for inshore fish species red drum and common snook to enable managers to begin incorporating environmental variables in management decisions. FWC managers are working with other agencies to assemble data on factors specific to each region—including harmful algal blooms, and seagrass, wetland, and mangrove coverage—to supplement standard stock assessments to inform regulations.

Great progress has been made by FWC staff to integrate habitat- and water quality–related factors into fisheries management decisions. However, these metrics address local factors (e.g., changes in seagrass coverage) and are reactive. The larger-scale stressors impacting local habitats (e.g., nutrient loads in freshwater flows, alteration of freshwater flow regimes, seagrass scarring by boat propellers) are not yet included. Therefore, the management structure permits fisheries regulations to be revised based on observed habitat metric values but does not influence management of the broader-scale factors affecting those habitat metrics. Improving Florida's fisheries management will require a landscape approach that accounts for connectivity at the watershed scale and integrates landscape and seascape in a more proactive fashion. This will require improved collaboration between the agencies responsible for managing Florida's water, habitats, land use, and fisheries and fully integrating their missions to enable holistic fisheries management.

The Status of Freshwater Springs Conservation and Watershed Management in Florida

Florida's freshwater plays an important role in fisheries management and presents a case study for land use and watershed protections within the state. Florida's freshwater is derived from a mixture of surface and subsurface sources, with a large portion provided by aquifers—Floridan, Biscayne, Sand and Gravel, Intermediate, and others that are unnamed. The largest is the Floridan aquifer system (FAS), which encompasses the entire state and portions of Georgia, South Carolina, and Alabama (Williams and Kuniansky 2015). Groundwater also serves as the primary source of potable water within the state for an estimated 18 million residents and a much larger nonresident tourist population (Marella 2020). The region where FAS recharge occurs covers about 110,000 km^2 (42,460 mi^2), or 65% of the state's land and water area (Knight and Meeks 2022). The unconfinement of the FAS is a defining characteristic of the region as millions of gallons of groundwater discharge from an estimated 1,113 freshwater artesian springs (Florida Geological Survey 2019). Thirty-three first-magnitude springs (defined as having a flow greater than 100 ft^3 per second or approximately 64.6 million gallons [245 million liters] of water per day) are documented in Florida, more than any other state or country (Rosenau et al. 1977; Scott et al. 2004). The unconfinement of the FAS, which allows for spring formation, is also the cause of its greatest vulnerabilities: contamination and depletion.

Springs help maintain ecological balance and support biodiversity in freshwater, coastal, and estuarine areas. Along the coast, freshwater inputs provided by spring flow act as a buffer against salinity changes in estuarine environments (Chen 2021) and thus influence estuarine ecology (Sklar and Browder 1998). Furthermore, springs provide habitat and food resources for a variety of species, including coastal fishes.

Freshwater springs provide critical hydrologic and climatic refugia on phenological and evolutionary time scales for both aquatic and terrestrial species (Cartwright et al. 2020; chapter 5, this book). This role will be increasingly important as the geographic ranges of tropical and subtropical species shift north due to climate change (Osland et al. 2021). For example, the spring-fed portions of rivers north of Tampa Bay maintain a relatively constant temperature year-round (~23°C), even as water temperatures near the Gulf of Mexico routinely drop below 10°C during winter. During cold periods, spring-fed rivers serve as thermal refugia for aggregations of cold-sensitive organisms such as manatees, which drive a thriving ecotourism industry (King and Heinen 2004), and subtropical fishes such as snook, which support recreational fisheries (Holzwart et al. 2022).

As is the case for coastal ecosystems, landscape management is poorly integrated with freshwater ecosystem conservation. Pollution, flow decline, and habitat loss in Florida's springs are largely driven by agricultural runoff and urban waste, water demands for residential and commercial use, and the physical alterations of spring habitats through management or excessive recreation. The direct impact of land-use practices on spring health has long been appreciated as heavy industrial groundwater use in the 1930s led to the cessation of flows at many springs, including Kissengen Spring (Lewelling et al. 1998) and White Sulphur Springs (Mirti 2001). This has facilitated a tenuous link between spring conservation efforts and the broader management of landscapes. The degradation of watersheds due to urbanization, agriculture, and industrial activities has led to increased nutrient pollution while limiting aquifer recharge. The culmination of these various stressors results in reduced biodiversity, decreased discharge, habitat loss, and altered water chemistry, affecting their ecological character and the broader estuarine environments they feed into. These stressors not only impact ecosystems but also affect the socioeconomic stability of local communities dependent on these ecosystems for recreation, tourism, and livelihoods such as fishing and shellfish harvesting (Borisova et al. 2014).

As population growth and land-use development have rapidly expanded, the impact on natural springs and their associated ecosystems has increased, necessitating concerted conservation efforts (North Florida Regional Water Supply Plan 2023). The most effective management practices have focused on preserving water quality and maintaining natural flow regimes essential for healthy aquatic ecosystems.

The management of watersheds involves defining and implementing strategies to maintain or restore natural hydrologic and ecological functions. Regulatory tools like Total Maximum Daily Loads (TMDLs) and Basin Management Action Plans (BMAPs) quantify pollutant loads and outline steps to meet water quality standards, protecting aquatic ecosystems linked to springs and estuaries. BMAPs are comprehensive programs designed to implement the necessary pollutant reductions specified by TMDLs, and involve local and state commitments, encompassing strategies like wastewater treatment improvements and agricultural Best Management Practices (BMPs). BMPs were developed to provide resources for practitioners on how best to manage their resources, including nutrient reduction, irrigation management, and protection of water resources while mitigating harm to the environment. Although BMPs are practical, cost-effective methods to manage pollutants and environmental impacts, they are voluntary, and application has been sporadic.

Another tool that can help protect springs and coastal areas is Florida's Minimum Flows and Levels (MFLs) law. MFLs refer to "the minimum flow for a watercourse or the minimum water level for groundwater in an aquifer or the minimum water level for a surface water body that is the limit at which further withdrawals would be significantly harmful to the water resources or ecology of the area." The goal is to set a protective hydrologic regime that will maintain the range of surface water fluctuations required to ensure healthy water resources, including species composition, vegetative structure, and ecological functions. MFLs have been set for roughly 30 Florida springs. Allowable average spring flow reductions in these MFLs range from 0–20%, with allowable flow decline due to human causes of 5–10% most common. An important component of MFLs is to maintain flows sufficient to support economically important fish species. For example, the MFL for the Homosassa River is based on providing snook with a thermal refuge during winter, and this metric will likely be applied to other rivers in the region (Herrick et al. 2019). However, a shifting historical baseline presents challenges to this approach. For example, it appears that reduction of freshwater flows from the Suwannee River has already caused a decline in blue crab abundance in coastal waters, which has resulted in a decline in the recreational fishery for tarpon (Burke 2021).

Maintaining and restoring hydrologic connectivity, water quality, and habitat integrity in Florida are the responsibility of a suite of regulatory and nonregulatory agencies. Florida's water management districts (WMDs; Suwannee River, St. Johns River, Southwest Florida, South Florida, and Northwest Florida), a network of four Florida National Estuary Programs (NEPs), and Aquatic Preserves managed by the Florida Department of Environmental Protection (FDEP) provide umbrellas for state and federal government agencies, local municipalities, and nongovernmental organizations to collaborate to address water and habitat management. However, fragmentation of management exacerbates the challenges faced by fish species that were outlined earlier. The governance of water resources is typically divided among several agencies (table 6.1).

Occasionally agencies have unified to address protection and restoration of Florida's springs. One major effort to protect and manage springs was the formation of the Florida Springs Protection Task Force (FSPTF) in 2014. The FSPTF was pivotal in assessing the environmental, social, and economic factors impacting spring basins across the state while educating the public on these issues. These efforts led to significant legislative and practical interventions, with one key success being development of BMAPs designed to reduce nitrogen pollution through comprehensive watershed-level management strategies. These plans involved local and state commitments to reduce pollutant loading and to outline specific projects and strategies for water quality restoration. An-

TABLE 6.1. Summary of the various Florida agencies responsible for the governance of water resources

MANAGEMENT ENTITY	DESCRIPTION
Florida Department of Environmental Protection (FDEP)	Oversees water quality and habitat protection statewide, but successful coordination with other local, state, and regional bodies, though critical for holistic management, is mixed.
Water management districts (WMDs)	Focus on regional water management but may lack sufficient integration with local land-use planning and wildlife and fisheries management agencies.
Municipalities	Responsible for local land-use decisions and wastewater treatment, which can directly impact nearby water bodies, but often lack the regulatory power to implement adequate guidance and restrictions and are rarely integrated with resource management agencies.
Florida Fish and Wildlife Conservation Commission (FWC)	Manages fisheries and wildlife but needs to better align with agencies that influence land and water use so that habitat conservation efforts are integrated with fisheries management to ensure sustainable practices are enacted.

other crucial element of the FSPTF's work was the designation of "Outstanding Florida Springs," which are springs of historical and ecological significance that require additional protection. The Florida legislature defined protections for 30 Outstanding Florida Springs in 2016, helping focus efforts and funding on these critical ecosystems. However, the FSPTF was disbanded, and the Blue-Green Algae Task Force (BGATF) was created in 2019 in response to extensive anthropogenic nutrient-fueled algae blooms throughout the state. The BGATF built upon the recommendations of the FSPTF and recommended immediate and extensive attention to addressing anthropogenic nutrient loads in Florida's waterways—including BMPs, BMAPs, TMDLs, Numeric Nutrient Criteria, and wastewater infrastructure—with consideration of impacts on downstream water bodies (FDEP 2019). Most of the BGATF recommendations have yet to be implemented. The integration of land-use planning with aquatic ecosystem conservation remains a significant challenge, necessitating enhanced collaborative frameworks that align local, regional, and state efforts to sustain the health and productivity of these critical ecosystems.

A Watershed Landscape Conservation Framework for Aquatic and Coastal Conservation in Florida

There is general agreement that successful conservation of Florida's aquatic ecosystems depends on clean water and habitat connectivity at local and regional scales. Florida has many of the components in place to enact integrated landscape approaches. However, current coastal conservation practices are piecemeal, and collaborative activities tend to be reactionary. For example, measures like habitat restoration and fishery closures do not address many of the causes of declines in ecosystems and the fisheries they support. The goal of this section is to propose a more holistic conservation strategy that integrates the landscape. It is based on Florida's terrestrial conservation approach focused on connectivity of habitat types and species pathways (see chapters 2, 3, 7, this book). Enacting this strategy is possible through better synergy between current conservation and management efforts with agencies, NGOs, and stakeholders, integration of new approaches, better mapping, and public education highlighting better coastal fisheries management.

Watershed conservation is at the core of a suite of approaches, including Landscape Conservation Design (LCD; Evans et al. 2019) and Integrated River Basin Management (IRBM; Pirrone et al. 2005). Integrated Land-Sea Management (ILSM) furthers involvement of coastal habitats by establishing linkages between IRBM and Integrated Coastal Zone Management (ICZM), enabling estuarine conservation at the watershed scale (Pirrone et al. 2005). Frequently, landscapes are assessed in isolation, despite the importance of landscape interconnectivity (Leonard et al. 2017), with some notable exceptions (the ridges to reefs approach on islands; Lagabrielle et al. 2011). Inclusion of coastal fishery habitats is rare.

We propose applying a systematic planning process to integrate landscapes (Leonard et al. 2017) with the overall goal of maintaining a sufficiently intact and connected watershed and coastal habitat mosaic network to support sustainable ecosystems, with a focus on coastal fishes.

Paramount to this framework is the focus on connectivity across landscapes—from watersheds to springs to estuaries and to the coastal habitat mosaic. Thus, both within and among landscapes, the focus should be on the important characteristics of habitat diversity (the number of different habitat types within a focal area) and habitat heterogeneity (the spatial arrangement among habitats), both of which comprise the habitat mosaic and influence connectivity (Foley et al. 2010). Connectivity includes surface and subsurface water flows (amount, timing, location, seasonality, quality) and fish and animal movements (ontogenetic habitat shifts, migrations, seasonal movements). The ultimate goal is to conserve or restore overall ecological function by focusing on protection and restoration of habitats and their connectivity.

Critical to this landscape conservation framework is collaboration. In Florida, this will require significant improvement in collaboration between the numerous agencies tasked with land, water, habitat, and marine fisheries management, NGOs, and stakeholders. Some such collaboration already occurs for terrestrial conservation—for example, the Florida Wildlife Corridor. A collaborative structure exists under the umbrella of NEPs for general estuary restoration and management, such as inclusion of municipal land-use planning in estuary management efforts, but this lacks mandates or enforcement. Additionally, stakeholder engagement is a standard component of natural resource management in Florida, but at present, these collaborations lack integration across landscapes and fail to incorporate fisheries. Creating and implementing this new collaborative structure will be the biggest challenge.

Here we outline a stepwise approach that follows established methods, as summarized above, that can be aggregated under the umbrella of systematic conservation planning.

STEP 1. IDENTIFY THE SPATIAL SCALE (WATERSHED) OF THE CONSERVATION ZONES OF FOCUS

In Florida, this is already somewhat defined by the WMDs, but these watershed designations should be reevaluated to ensure they address ecological connectivity that incorporates coastal systems. This has been done to a large extent for Florida's four NEP estuaries, which provide a good template. However, mosquito ditches and water conveyance canals/ditches used to manage freshwater flows for flood management and agriculture frequently cross watershed boundaries and are generally not considered in watershed delineation. In addition, watersheds within WMDs differ in spatial scale—some are small watersheds that connect to tidal creeks, whereas others are river-scale watersheds. The former directly connect to nursery and adjacent habitats for economically important species like snook and tarpon (as well as a creek-dependent fish community), which are impacted by changes in local land use (Wilson et al. 2019, 2023). The latter are more likely to have effects at the estuary scale via changes in freshwater flow quality, amount, quantity, and timing (Sklar and Browder 1998).

STEP 2. MAP LAND USE WITHIN EACH CONSERVATION ZONE

Much of these data are already available, though databases are often held within different agencies and are organized for different purposes. It will be essential to construct databases for each watershed that include all components of land use, such as ownership (government vs. private), zoning (agriculture vs. urban vs. conservation), planning (municipal development or conservation plans), projected land use, water management (flood control, freshwater withdraws, aquifer vs. surface water), and connectivity to coastal habitats. Within this mapping, future land-use alternatives should be considered (see the section "Engage with Stakeholders"), with regular updates.

STEP 3. USE A DRIVER-PRESSURE-STATE-IMPACT-RESPONSE (DPSIR) STRUCTURE TO ASSESS THE STATUS OF THE WATERSHED AND INFORM CONSERVATION PLAN DESIGN (E.G., GILBY ET AL. 2023)

These tasks should not only address localized watershed characteristics, but also downstream effects on the estuarine or coastal component of the conservation zone and the coastal habitat mosaic. In summary: drivers are processes and anthropogenic activities that cause pressures; pressures are the direct environmental stresses caused by the drivers; impact is the measure of the effects of the drivers and pressures on the state of environmental system (impacts are in part measured by assessing the environmental state of the natural systems); response is the evaluation of possible actions to enact as management measures. In many cases, data are available to enable these assessments, but sometimes data will need to be collected or reexamined in the broader context of watershed-coastal conservation. Expert elicitation can be used to obtain data via a multicriteria decision-making process (Margules and Sarkar 2007). Threats can be evaluated through a vulnerability analysis that integrates spatial habitat and species data and uses expert assessments to determine vulnerability to anthropogenic threats (Foley et al. 2010). The impact metric of the DPSIR approach should be focused on the coastal zone, which is traced to the drivers within the watershed (Pirrone et al. 2005) that need to be addressed (response) as part of a landscape conservation framework, similar to the approach used to guide land use and conservation on tropical islands, with a focus on coral reef health (e.g., Lagabrielle et al. 2011). This requires integrating different landscapes (watershed and estuarine/marine).

Importantly, the influence of current and future anthropogenic activities needs to be assessed within the context of historical conditions. On the one hand, returning an ecosystem in Florida to the environmental state that it was 150 years ago is highly unlikely, and it is important to use modern conditions to ensure that management recommendations are relevant to current conditions or desired future conservation outcomes rather than distant historical baselines (Gilby et al. 2022). However, the historical state of watersheds is useful as a benchmark to guide restoration and conservation efforts. For example, Lawlor et al. (2023) mapped predevelopment wetland deepwater habitats in St. Lucie County, Florida, to present baseline conditions that allow projects to better mitigate for changes in land cover due to development.

Similarly, conservation plans should focus on multiple future scenarios to maximize conservation effectiveness and incorporate an adaptive structure within the plan. For example, the 1000 Friends of Florida maps of land use provide a 2019 baseline, and project land-use outcomes in 2040 under sprawl and conservation scenarios (1000 Friends of Florida undated; see chapter 4). A watershed-coastal landscape conservation plan for Florida should aim to achieve the 2040 Conservation projection but also create alternatives for a sprawl scenario. This structure will be instrumental for reaching a consensus on a preferred vision with public support, political will, and funding.

STEP 4. ENGAGE WITH STAKEHOLDERS

Stakeholders in the conservation planning zone should be incorporated into a coproduction process. This is a standard component of Landscape Conservation Design (Evans et al. 2019), a process used extensively by the US Fish and Wildlife Service. In Florida, the Rural and Family Lands Protection Program is already used to integrate stakeholders (residents, property owners) in conservation efforts (see chapters 3 and 9), and stakeholders in fisheries, hunting, ecotourism, and other uses are typically included in fisheries and wildlife management planning via working groups or public workshops.

For watershed conservation, some agricultural uses should be considered the best use of land relative to residential/urban and commercial development, especially when agricultural landowners follow effective and evaluated BMPs. Here, the efforts of the Healthy Farms Healthy Bays (HFHB) program should be consulted (Solutions from the Land undated). This program, focused on the Suwannee River region, seeks to identify ongoing conservation programs within the region; identify and collaborate with stakeholders and existing groups involved in conservation; identify and access available cost-share programs between agencies and others to promote water conservation; and integrate consideration of downstream impacts to coastal areas from land-use practices. Similarly, the Floridan Aquifer Collaborative Engagement for Sustainability (FACETS) aims to achieve agricultural water security while meeting environmental standards by managing water and land-use patterns, though there is no explicit inclusion or integration of estuaries downstream of focal areas. The key with FACETS will be to ensure the different landscapes of watersheds and estuaries are integrated, as in the HFHB initiative.

Design and implementation of landscape conservation, as proposed here, may cause socioeconomic changes that reduce broad acceptance. Therefore, as part of this process a cost-benefit analysis to compare the socioeconomic benefits from conservation and ecosystem services against other land-use options should be conducted (Lagabrielle et al. 2011). Similarly, the socioeconomic costs and benefits of different conservation scenarios should be evaluated. As part of the socioeconomic evaluation, the different temporal scales of economic measures must be considered: In general, most economic measures are short-term, whereas the environmental and economic measures of conservation are typically long-term. For example, long-term conservation costs include costs of plan implementation, conservation management, restoration (if needed), and transformation pressure, such as urbanization adjacent to conservation lands and invasive plants (Lagabrielle et al. 2011).

STEP 5. CREATE AND EVALUATE A PLANNING FRAMEWORK

With conservation zones identified, data in hand, data gaps known, and stakeholders engaged, use an interactive, iterative process to evaluate and prioritize conservation scenarios for each watershed-estuary conservation zone. This should include setting explicit conservation targets and defining what constitutes adequate protection so that performance can be evaluated in the future and adjustments made if needed. Of particular importance are foundation species, which are species that make a large contribution toward creating a habitat/

ecosystem that supports other species (Osland et al. 2022). Foundation species in marine coastal areas include mangroves, seagrasses, oysters, and salt marsh plants. Since a focus of this approach is estuary and fish conservation, conservation targets should include estuarine habitats and target fish species. For example, juvenile snook and tarpon require a habitat mosaic of mangrove and marsh ponds, creeks, wetlands, and shorelines, with connectivity to freshwater areas (Barbour et al. 2014; Bunting et al. 2024). Thus, conservation targets should include maintaining the integrity of the habitat mosaic with snook and tarpon as surrogate species for broader conservation benefits and prioritizing locations for protection and restoration (Wilson et al. 2023).

Alteration of freshwater flows into mangrove and marsh habitats negatively impacts species composition, which decreases prey availability and diet for juvenile snook, so restoration of altered freshwater flow patterns and maintenance of natural flow patterns should also be a conservation target. Similarly, freshwater flows and nutrient loads that exacerbate plankton blooms and cause seagrass declines should be conservation targets, since loss of seagrass negatively affects populations of important recreational fish species like spotted seatrout and red drum. At present, some of these themes are being addressed (e.g., via BMAPs and TMDLs) but only on general scales, typically without enforcement and without inclusion of fish or fisheries as the ultimate conservation targets or fish and habitat metrics that can be used at the landscape scale.

Systematic conservation planning is an iterative, interactive, adaptive process for mapping conservation area networks to maximize complementarity among areas (Margules and Sarkar 2007). Given the inherent interactions between conservation zones (between watersheds, between watersheds and estuarine areas, between estuarine areas), a systematic conservation planning framework should be used to optimize conservation action outcomes that account for connectivity between zones. This process will need to take into consideration the locations, spatial extent, and status of existing conservation zones to maximize zone connectivity and synergy.

Conclusion and Additional Considerations

The approach proposed here is not novel; indeed, similar approaches have previously been proposed (e.g., DeFreese 1995), and it is subject to the challenges inherent in collaborative network governance (see Bixler et al. 2016 and articles within the associated special issue). However, implementation will be a challenge because the extensive collaborative structure needed for landscape watershed-to-estuary conservation to succeed is not currently in place. Indeed, the authors are aware of numerous initiatives focused on watershed restoration and conservation, but these programs do not incorporate estuarine/coastal landscapes, the coastal habitat mosaic, or coastal fish species. Thus, the conservation approach we outline here will require a paradigm shift.

The underlying assumption of this landscape conservation approach is that we are targeting habitat mosaics as our proxy for biodiversity and ecological function. Therefore, habitat type, heterogeneity, and connectivity should be used as core environmental variables for structuring conservation plans. Habitat characteristics can be used as proxies for biodiversity (Foley et al. 2010), and data on habitats and land use are easier to obtain than species data. A habitat-based approach may also address funding/personnel limitations at agencies, which often don't have resources to maintain species databases. Although a species-specific approach, especially using multiple focal species, is more effective for terrestrial conservation connectivity than a "naturalness" or habitat-focused approach (Krosby et al. 2015), species-based approaches tend to be more data-intensive and time consuming, especially since focal species may differ among conservation zones. In addition, in terrestrial work, the number and types of surrogate species needed also change with land use (Dutta et al. 2023).

Uncertainties in the dynamic interactions between the watershed and coastal zone dictate a precautionary approach, which provides a buffer for estimates of conservation needs and benefits and allows for better adaptation to environmental changes (Foley et al. 2010). Discomfort with uncertainty is a big concern with marine fisheries management. For example, fisheries managers want to know exactly how many fish a nursery habitat will produce, but this is unfeasible due to the high variability between habitats and variable fish recruitment among years. This underscores the foundational philosophy behind the landscape habitat-based approach—sufficient healthy and connected habitats create a resilient system that provides sufficient fish for a fishery, even if we're unable to discern an exact number of fish produced per annum.

To a great extent, the most effective way to build ecosystem resilience to climate change is to address the stressors (excess nutrients, contaminants, failed infrastructure) that compromise the ecological integrity of these systems (Parkinson et al. 2021), which is the focus of the approach proposed here. However, an adaptive approach is mandatory to account for the uncertainties from climate change effects by including Resist-Accept-Direct (RAD) strategies that address climate change (e.g., Williams and Brown 2024). For example, sea-level rise will create pressures to shift estuarine wetlands inland, but in many locations, development has precluded the ability of wetlands to migrate inland. In areas where wetlands can migrate, freshwater habitats will be displaced. Sea-level rise will also cause saltwater intrusion, exacerbated by anthropogenic water withdrawals. Changing precipitation patterns will likely increase the occurrence of extreme droughts and extreme rainfall events, which will interact with anthropogenic influences like impervious surfaces and water control infrastructure and have both local (within watersheds) and downstream (in estuaries) ecosystem-level effects (Walker et al. 2021).

Finally, this new paradigm must tackle numerous systemic challenges, including

- outdated and inadequate wastewater and stormwater infrastructure, which is being exacerbated by continued rapid population growth with inadequate policies and funding mechanisms;
- lack of a landscape-level vision for a preferred future, which is inhibited by a fragmented regulatory system; and
- inadequate funding for monitoring, mapping, and modeling at appropriate landscape-level spatial and temporal scales.

Literature Cited

1000 Friends of Florida. Undated. Florida's Rising Seas: Mapping Our Future. http://1000fof.org/sealevel2040/sea-level-2040/#state2040. Accessed August 18, 2024.

Adams, A. J., R. K. Wolfe, and C. A. Layman. 2009. Preliminary examination of how human-driven freshwater flow alteration affects trophic ecology of juvenile snook (*Centropomus undecimalis*) in estuarine creeks. Estuaries and Coasts 32:819–828.

Barbour, A. B., A. J. Adams, and K. Lorenzen. 2014. Size-based, seasonal, and multidirectional movements of an estuarine fish species in a habitat mosaic. Marine Ecology Progress Series, 507:263–276.

Bixler R. P., M. McKinney, and L. Scarlett. 2016. Forging new models of natural resource governance. Frontiers in Ecology and the Environment 14:115. https://doi.org/10.1002/fee.1255.

Borisova, T. A., W. Hodges, and T. J. Stevens. 2014. Economic Contributions and Ecosystem Services of Springs in the Lower Suwannee and Santa Fe River Basins of North-Central Florida. University of Florida, Food and Resource Economics Department, Gainesville.

Boucek, R. E., M. S. Allen, R. D. Ellis, J. Estes, S. Lowerre-Barbieri, and A. J. Adams. 2023. An extreme climate event and extensive habitat alterations cause a non-linear and persistent decline to a well-managed estuarine fishery. Environmental Biology of Fishes 106:193–207.

Bunting, M. S., P. W. Stevens, D. A. Blewett, Q. M. Tuckett, J. E. Hill, G. R. Poulakis, C. P. Shea, and C. R. Saari. 2024. Emigration of juvenile tarpon *Megalops atlanticus* from ephemerally connected coastal ponds. Estuaries and Coasts 47:2493–2509.

Burke, M. 2021. Lords of the Fly: Madness, Obsession, and the Hunt for the World Record Tarpon. Pegasus Books, Cambridge, UK.

Caddy, J. F. 2000. Marine catchment basin effects versus impacts of fisheries on semi-enclosed seas. ICES Journal of Marine Science 57:628–640.

Cartwright, J. M., K. A. Dwire, Z. Freed, S. J. Hammer, B. McLaughlin, L. W. Misztal, E. R. Schenk, J. R. Spence, A. E. Springer, and L. E. Stevens. 2020. Oases of the future? Springs as potential hydrologic refugia in drying climates. Frontiers in Ecology and the Environment 18:243–253.

Castillo, N. A., R. O. Santos, W. R. James, R. Rezek, D. Cerveny, R. Boucek, A. J. Adams, J. Fick, T. Brodin, and J. Rehage. 2023. Widespread pharmaceutical exposure at concentrations of concern for a subtropical coastal fishery: Bonefish (*Albula vulpes*). https://dx.doi.org/10.2139/ssrn.4394268.

Chen, X. 2021. Hydrodynamic modeling of salinity habitat changes with reduced submarine groundwater discharges in a spring-fed estuary. Regional Studies in Marine Science 43(March): 101674. https://doi.org/10.1016/j.rsma.2021.101674.

De Freese, D. E. 1995. Land acquisition: A tool for biological diversity protection in the Indian River Lagoon, Florida. Bulletin of Marine Science 57(1):14–27.

Duarte, C. M., D. J. Conley, J. Carstensen, and M. Sánchez-Camacho. 2009. Return to Neverland: Shifting baselines affect eutrophication restoration targets. Estuaries and Coasts 32:29–36. https://doi.org/10.1007/s12237-008-9111-2.

Dutta, T., M. De Barba, N. Selva, A. C. Fedorca, L. Maiorano, W. Thuiller, A. Zedrosser, J. Signer, F. Pfluger, S. Frank, P. M. Lucas, and N. Balkenhol. 2023. An objective approach to select surrogate species for connectivity conservation. Frontiers in Ecology and Evolution. https://doi.org/10.3389/fevo.2023.1078649.

Evans, N. M., A. L. Carrozzino-Lyon, B. Galbraith, J. Noordyk, D. M. Peroff, J. Stoll, A. Thompson, M. W. Winden, and M. A. Davis. 2019. Ecosystem Services 39:101001. https://doi.org/10.1016/j.ecoser.2019.101001.

Ferreira, G. V. B., M. Barletta, and A. R. A. Lima. 2019. Dynamics of marine debris ingestion by profitable fishes along the estuarine ecocline. Scientific Reports 9:13514. https://doi.org/10.1038/s41598-019-49992-3.

Florida Department of Environmental Protection (FDEP). 2023. Blue-Green Algae Task Force Final Consensus #1. https://floridadep.gov/sites/default/files/Final%20Consensus%20%231_0.pdf. Accessed June 28, 2024.

Florida Fish and Wildlife Conservation Commission (FWC). 2024a. Red tide closures https://myfwc.com/fishing/saltwater/rulemaking/history/. Accessed August 18, 2024.
Florida Fish and Wildlife Conservation Commission (FWC). 2024b. https://myfwc.com/fishing/saltwater/recreational/red-drum/. Accessed August 18, 2024.
Florida Geological Survey, Florida Springs [map]. January 30, 2019. FDEP Open Data Portal, Florida Department of Environmental Protection. https://geodata.dep.state.fl.us/datasets/FDEP::florida-springs/explore?location=28.564918%2C-84.175450%2C6.78. Accessed June 25, 2024.
Foley, M. F., B. S. Halpern, F. Micheli, M. H. Armsby, M. R. Caldwell, C. M. Crain, E. Prahler, N. Rohr, D. Sivas, M. W. Beck, M. H. Carr, L. B. Crowder, J. E. Duffy, S. D. Hacker, K. L. McLeod, S. R. Palumbi, C. H. Peterson, H. M. Regan, M. H. Ruckelshaus, P. A. Sandifer, and R. S. Steneck. 2010. Guiding ecological principles for marine spatial planning. Marine Policy 34:955–966.
Fredley, J., M. J. Durako, and M. O. Hall. 2019. Multivariate analyses link macrophyte and water quality indicators to seagrass die-off in Florida Bay. Ecological Indicators 101:692–701.
Gilby, B. L., L. A. Goodridge Gaines, H. P. Borland, C. J. Henderson, J. D. Mosman, A. D. Olds, and H. J. Perry. 2023. Drivers of ecological condition identify bright spots and sites for management across coastal seascapes. Estuaries and Coasts 46:906–924. https://doi.org/10.1007/s12237-023-01187-w.
Hagy, J. D., W. R. Boynton, C. W. Keefe, and K. V. Wood. 2004. Hypoxia in Chesapeake Bay, 1950–2001: Long-term change in relation to nutrient loading and river flow. Estuaries 27:634–658.
Heil, C. A., and A. L. Muni-Morgan. 2021. Florida's harmful algal bloom (HAB) problem: Escalating risks to human, environmental and economic health with climate change. Frontiers in Ecology and Evolution 9:646080.
Herrick, G., X. Chen, C. Anastasiou, R. Basso, N. Mendez-Ferrer, N. Ortega, D. Rogers, and D. A. Leeper. 2019. Reevaluation of minimum flows for the Homosassa River System. Southwest Florida Water Management District. https://www.swfwmd.state.fl.us/sites/default/files/documents-and-reports/reports/Hom%20MFL%20FINAL%20Oct%202019.pdf. Accessed August 18, 2024.
Holzwart, K. R., K. Deak, J. Miller, E. Johnson, L. Simonton, T. Dluzniewski, A. Stanfill, and A. Taylor. 2022. Common Snook (*Centropomus undecimalis*) establishment in and use of Nature Coast springs systems. Florida Scientist 85:26–43.
Jung, N. W., G. H. Lee, T. M. Dellapenna, Y. Jung, T. C. Jo, J. Chang, and S. M. Figueroa. 2024. Economic development drives massive global estuarine loss in the Anthropocene. Earth's Future 12(4), e2023EF003691.
King, J. M., and J. T. Heinen. 2004. An assessment of the behaviors of overwintering manatees as influenced by interactions with tourists at two sites in Central Florida. Biological Conservation 117:227–234.

Knight, R. L., and A. Meeks. 2022. Assessing pollution and depletion of large artesian springs in Florida's rapidly developing water-rich landscape. Pages 7–18 in M. J. Currell and B. G. Katz, editors. Threats to Springs in a Changing World. American Geophysical Union (AGU). https://doi.org/10.1002/9781119818625.ch2.

Krosby, M., I. Breckheimer, D. J. Pierce, P. H. Singleton, S. A. Hall, K. C. Halupka, W. L. Gaines, R. A. Long, B. H. McRae, B. L. Cosentino, and J. P. Schuett-Hames. 2015. Focal species and landscape "naturalness" corridor models offer complementary approaches for connectivity conservation planning. Landscape Ecology 30:2121–2132.

Lagabrielle, E., M. Rouget, T. Le Bourgeois, K. Payet, L. Durieux, S. Baret, J. Dupont, and D. Strasberg. 2011. Integrating conservation, restoration and land-use planning in islands—An illustrative case study in Réunion Island (Western Indian Ocean). Landscape and Urban Planning. dx.doi.org/10.1016/j.landurbplan.2011.02.004.

Lawlor, S. K., S. Raines, M. Landry, and M. Rains. 2023. Forensic wetland and deepwater habitat mapping for setting pre-development conditions. Wetland Science and Practice 2023:234–240.

Leonard, P. B., R. F. Baldwin, R. D. Hanks. 2017. Landscape-scale conservation design across biotic realms: Sequential integration of aquatic and terrestrial landscapes. Scientific Reports 7:14556.

Lewelling, B. R., A. B. Tihansky, and J. L. Kindinger. 1998. Assessment of the hydraulic connection between ground water and the Peace River, west-central Florida. 97–4211. Water-Resources Investigations Report. US Geological Survey. https://doi.org/10.3133/wri974211.

Lewis, D. M., and G. S. Cook. 2023. Freshwater discharge disrupts linkages between the environment and estuarine fish community. Ecological Indicators 151:110282.

Marella, R. L. 2020. Water withdrawals, uses, and trends in Florida, 2015. 2019–5147. US Geological Survey. https://doi.org/10.3133/sir20195147.

Margules, C., and S. Sarkar. 2007. Systematic Conservation Planning. Cambridge University Press, Cambridge, UK.

Medina, M., D. Kaplan, E. C. Milbrandt, D. Tomasko, R. Huffaker, and C. Angelini. 2022. Nitrogen-enriched discharges from a highly managed watershed intensify red tide (*Karenia brevis*) blooms in southwest Florida. Science of the Total Environment 827:154149.

Mirti, T. H. 2001. Springflow assessment of White Sulphur Springs. Georgia Institute of Technology. http://hdl.handle.net/1853/44453.

Morris, L. J., L. M. Hall, C. A. Jacoby, R. H. Chamberlain, M. D. Hanisak, J. D. Miller, and R. W. Virnstein. 2022. Seagrass in a changing estuary, the Indian River Lagoon, Florida, United States. Frontiers in Marine Science 8:789818.

North Florida Regional Water Supply Plan. 2023. https://northfloridawater.com/watersupplyplan/index.html.

O'Leary, J. K., M. C. Goodman, R. K. Walter, K. Willits, D. J. Pondella, and J. Stephens.

2021. Effects of estuary-wide seagrass loss on fish populations. Estuaries and Coasts 44:2250–2264.
Olin, J. A., P. W. Stevens, S. A. Rush, N. E. Hussey, and A. T. Fisk. 2015. Loss of seasonal variability in nekton community structure in a tidal river: Evidence for homogenization in a flow-altered system. Hydrobiologia 744:271–286.
Osland, M. J., A. R. Hughes, A. R. Armitage, S. B. Scyphers, J. Cebrian, S. H. Swinea, C. C. Shepard, M. S. Allen, L. C. Feher, J. A. Nelson, and C. L. O'Brien. 2022. The impacts of mangrove range expansion on wetland ecosystem services in the southeastern United States: Current understanding, knowledge gaps, and emerging research needs. Global Change Biology 28:3163–3187.
Osland, M. J., P. W. Stevens, M. M. Lamont, R. C. Brusca, K. M. Hart, J. H. Waddle, C. A. Langtimm, C. M. Williams, B. D. Keim, A. J. Terando, E. A. Reyier, K. E. Marshall, M. E. Loik, R. E. Boucek, A. B. Lewis, and J. A. Seminoff. 2021. Tropicalization of temperate ecosystems in North America: The northward range expansion of tropical organisms in response to warming winter temperatures. Global Change Biology 27(13): 3009–3034.
Parkinson, R. W., V. Seidel, C. Henderson, and D. E. DeFreese. 2021. Adaptation actions to reduce impairment of Indian River Lagoon water quality caused by climate change. Journal of Coastal Management 49:215–232. https://doi.org/10.1080/08920753.2021.1875399.
Phlips, E. J., S. Badylak, A. L. Mathews, E. C. Milbrandt, L. R. Montefiore, E. S. Morrison, N. Nelson, and B. Stelling. 2023. Algal blooms in a river-dominated estuary and nearshore region of Florida, USA: The influence of regulated discharges from water control structures on hydrologic and nutrient conditions. Hydrobiologia 850:4385–4411.
Pirrone, N., G. Trombino, S. Cinnirella, A. Algieri, G. Bendoricchio, and L. Palmeri. 2005. The Driver-Pressure-State-Impact-Response (DPSIR) approach for integrated catchment-coastal zone management: Preliminary application to the Po catchment-Adriatic Sea coastal zone system. Regional Environmental Change 5:111–137. https://doi.org/10.1007/s10113-004-0092-9.
Polidoro, B. A., K. E. Carpenter, L. Collins, N. C. Duke, A. M. Ellison, J. C. Ellison, E. J. Farnsworth, E. S. Fernando, K. Kathiresan, N. E. Koedam, and S. R. Livingstone. 2010. The loss of species: Mangrove extinction risk and geographic areas of global concern. PLOS One 5(4): e10095.
Pulster, E. L., K. Rullo, S. Gilbert, T. M. Ash, B. Goetting, K. Campbell, S. Markham, and S. A. Murawski. 2022. Assessing per-and polyfluoroalkyl substances (PFAS) in sediments and fishes in a large, urbanized estuary and the potential human health implications. Frontiers in Marine Science 9:1046667.
Rosenau, J. C., G. L. Faulkner, C. W. Hendry Jr., and R. W. Hull. 1977. Springs of Florida. Bulletin No. 31. Florida Geological Survey, Tallahassee.

Sklar, F. H., and J. A. Browder. 1998. Coastal environmental impacts brought about by alterations in freshwater flow in the Gulf of Mexico. Environmental Management 22:547–562. https://doi.org/10.1007/S002679900127.

Solutions from the Land. Undated. Florida's Healthy Farms, Healthy Bays project. https://www.solutionsfromtheland.org/log/floridas-healthy-farms-healthy-bays-project-aims-to-foster-farmer-fishermen-collaboration-for-cleaner-water/. Accessed August 18, 2024.

Stamp, T., E. West, T. Robbins, S. Plenty, and E. Sheehan. 2022. Large-scale historic habitat loss in estuaries and its implications for commercial and recreational fin fisheries. ICES Journal of Marine Science 79:1981–1991.

Walker, L. M., P. A. Montagna, X. Hu, and M. S. Wetz. 2021. Timescales and magnitude of water quality change in three Texas estuaries induced by passage of Hurricane Harvey. Estuaries and Coasts 44:960–971.

Walsh, C. J., A. H. Roy, J. W. Feminella, P. D. Cottingham, P. M. Groffman, and R. Morgan. 2005. The urban stream syndrome: Current knowledge and the search for a cure. Journal of the North American Benthological Society 24:706–723.

Williams, B. K., and E. D. Brown. 2024. Managing ecosystems with resist-accept-direct (RAD). Methods in Ecology and Evolution 15:796–805. https://doi.org/10.1111/2041-210X.14309.

Williams, L. J., and E. L. Kuniansky. 2015. Revised hydrogeologic framework of the Floridian Aquifer System in Florida and parts of Georgia, Alabama, and South Carolina. 1807. Professional Paper. US Geological Survey. https://doi.org/10.3133/pp1807.

Wilson, J. K., A. J. Adams, and R. N. Ahrens. 2019. Atlantic tarpon (*Megalops atlanticus*) nursery habitats: Evaluation of habitat quality and broad-scale habitat identification. Environmental Biology of Fishes 102:383–402.

Wilson, J. K., P. W. Stevens, D. A. Blewett, R. Boucek, and A. J. Adams. 2023. A new approach to define an economically important fish as an umbrella flagship species to enhance collaborative stakeholder-management agency habitat conservation. Environmental Biology of Fishes 106:237–254.

Woo, I., M. J. Davis, C. S. Ellings, S. Hodgson, J. Y. Takekawa, G. Nakai, and S. E. W. De La Cruz. 2019. A mosaic of estuarine habitat types with prey resources from multiple environmental strata supports a diversified foraging portfolio for juvenile chinook salmon. Estuaries and Coasts 42:1938–1954.

Yee, S. H., J. F. Carriger, P. Bradley, W. S. Fisher, and B. Dyson. 2015. Developing scientific information to support decisions for sustainable coral reef ecosystem services. Ecological Economics 115:39–50.

7

The Florida Ecological Greenways Network and Florida Wildlife Corridor

Protecting Biodiversity and Ecosystem Services

Thomas S. Hoctor, Michael O'Brien, Carlton Ward Jr., Mallory Dimmitt, Joseph M. Guthrie, Eve Bohnett, Sarah Lockhart, Asena Goren, Reed F. Noss, Jason Lauritsen, Marly Fuller, Kailyn Lawson, and Derya Yesilkusak

Introduction

Since the 1980s, Florida has been a leader in science-based corridor planning, ultimately implementing a comprehensive strategy to address habitat loss and fragmentation. The Florida Ecological Greenways Network (FEGN), established in 1997, is the ecological cornerstone of the state's greenways initiative. This chapter examines the FEGN's history, emphasizing its role in land-use planning and conservation. It also examines the impact of the FEGN on protecting biodiversity and ecosystem services, which resulted in establishing the Florida Wildlife Corridor, a statewide wildlife corridor plan.

Historical Context and Early Development

As conservation biology and landscape ecology emerged in the late 1970s and early 1980s, habitat loss and fragmentation became a concern to many scientists and conservationists in Florida. Larry Harris and Reed Noss, researchers at the University of Florida, laid the conceptual foundations of the FEGN in the 1980s by producing maps showing desired "landscape linkages" between particular conservation areas within the state and to adjacent states (e.g., Okefenokee National Wildlife Refuge; Noss and Harris 1986), with Noss producing a statewide Florida Wildlife Corridor design in 1985 (see chapter 2). The work of Harris and Noss served as a catalyst for discussions and action in the 1990s by conservation groups and Florida government on conservation strategies to mitigate the effects of rapid human population growth and habitat loss. This dialogue ultimately led to the establishment of the Florida Greenways Program and the FEGN. This forward-thinking approach enabled Florida to become a national leader in protecting biodiversity and ecosystem services and establishing innovative rural land conservation partnerships despite the state's ongoing rapid population growth.

Evolution and Impact of the FEGN: Modeling, Prioritization, and Integration

The FEGN originated as the ecological component of the Florida Greenways Plan, following the recommendations of the Florida Greenways Commission in 1994 (see chapter 3). Developed through collaboration between the University of Florida and the Florida Department of Environmental Protection and supported by the Florida Department of Transportation (FDOT) and US Department of Transportation, the FEGN's early work began in 1995. A team from the University of Florida led by Peggy Carr, Paul Zwick, and Tom Hoctor used GIS technology and comprehensive geographic data to identify trails and ecological greenways for the proposed Florida Greenways Program, involving extensive research, outreach, and public engagement and leading to new conservation legislation to enact a state greenways program.

Tom Hoctor led the development of the original FEGN model and boundaries, working with a large team of science advisers named the technical advisory group (TAG). From 1995 to 1997, the TAG helped refine data, criteria, and modeling methods to shape the FEGN's first delineation. The FEGN aimed to connect large, intact natural and rural landscapes into a statewide conservation system, providing ecological linkages and wildlife corridors (Hoctor et al. 2000, 2001; Hoctor 2003; Hoctor et al. 2004). Key goals included protecting habitats for viable populations of the Florida panther (*Puma concolor*) and Florida black bear (*Ursus americanus floridanus*), conserving other fragmentation-sensitive

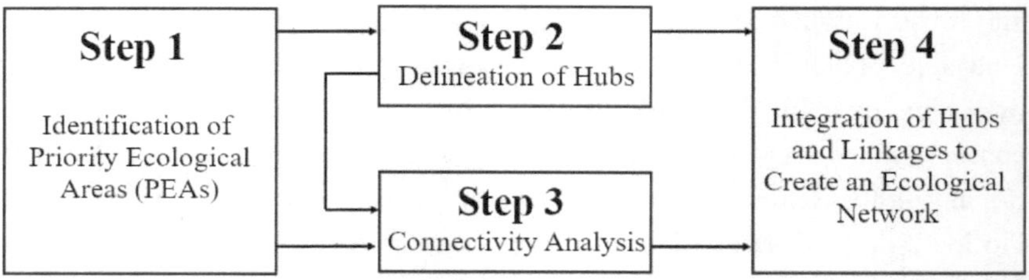

FIGURE 7.1. The process to delineate the FEGN boundary.

species, maintaining natural communities, and sustaining ecosystem services like freshwater resources, storm buffering, flood storage, climate resilience, and carbon sequestration.

The initial edition of the FEGN, finalized in 1997, was founded on a growing wealth of Florida GIS data. However, as with all conservation plans, the FEGN required continual revisions to maintain relevance by keeping up with changes in land-use and ecological data. As conservation needs and technologies advanced, the FEGN underwent significant updates in 2013 (Hoctor et al. unpublished) and 2021 (Hoctor et al. 2025), with another update in 2025. These revisions reflect advancements in scientific methods and data capabilities, ensuring the FEGN remains a vital tool for conservation planning in Florida.

SUMMARY OF THE FEGN BOUNDARY DELINEATION PROCESS

There are four primary parts of the FEGN boundary delineation process:

1. Identification of areas of ecological significance, termed Priority Ecological Areas (PEAs)
2. Delineation of hubs that encompass large areas of PEAs
3. Connectivity analysis to identify linkages between hubs
4. Creation of a statewide ecological network based on the integration of the hubs and connecting linkages identified in the analyses (fig. 7.1)

Florida was a leader in the creation of several data resources for conservation in the early 1990s, as shown by the Florida Fish and Wildlife Conservation Commission's *Closing the Gaps* report (Cox et al. 1994) and the Florida Natural Areas Inventory (FNAI) Potential Natural Areas (FNAI 2024a), both resources

TABLE 7.1. Priority Ecological Area criteria from the 2021 FEGN

FEGN PEA LAYERS	SOURCE
Landscape Species Models	FNAI, FWC, UF
Matrix-Landscape Natural Communities	CLC version 3.4, FNAI modified version
Florida Panther Conservation Zones	USFWS
Priority Panther Habitat	USFWS
Priority Florida Black Bear Habitat	University of Maryland and FWC
Existing Conservation Lands	FNAI FLMA
Strategic Habitat Conservation Areas	FWC
Rare Species Habitat Conservation Priorities	FNAI
FL Forever Underrepresented Natural Communities	FNAI
FL Forever Functional Wetlands	FNAI
FL Forever Natural Floodplains	FNAI
Potential Natural Areas (PNAs)	FNAI
Coastal Barrier Resources Act lands	DEP
Dark Sky Regions	Heimholtz Centre
Air Force Conservation Partnership Priority Focal Species Habitat	US Air Force, USFWS, UF, FNAI, FWC

serving as foundations of the FEGN delineation process. Since the inception of the Florida Forever conservation land protection program in 2000, FNAI has collaborated with numerous partners, including the University of Florida, to develop, revise, and enhance the Florida Forever Needs Assessment (FNAI 2023; see chapter 3). Furthermore, FNAI and the University of Florida have established and refined the Florida Critical Lands and Waters Identification Project (CLIP) since 2006, which delineates conservation priorities for biodiversity, water resources, and landscapes (Oetting et al. 2016). Comprehensive data sources on conservation priority areas have consistently enhanced the information used to establish PEAs across the multiple FEGN revisions from 1997 to 2021 with a variety of data vetted by the TAG that identify state biodiversity and ecosystem conservation priority areas (table 7.1).

Hubs are large clusters of PEAs and other compatible rural lands identified through spatial optimization to buffer and enlarge these conservation priority areas. The minimum size for hubs has been set at 2,000 ha (5,000 ac) since the FEGN's inception, after much deliberation by the original FEGN TAG regarding a size threshold that could best represent landscape-scale ecosystem protection (Hoctor et al. 2000). Subsequent TAGs for all the updates have maintained this size threshold as part of the criteria in all major updates.

Connectivity models have grown increasingly sophisticated since FEGN's initial release. The original 1997 model could only identify linkages through suitable habitats between hubs, but by 2013, the process had expanded to include seven distinct connectivity models:

1. The riverine/riparian corridor buffers and connectivity model applied an 800 m buffer to the 50 Major Rivers and the intersecting Outstanding Florida Waters (FREAC 1990). These buffers encompass adjacent compatible land uses, including both natural and seminatural areas, to safeguard water quality and hydrological processes. Additionally, they support habitat gradients and wildlife movement corridors along rivers.
2. The xeric natural community connectivity model connects xeric habitats, such as upland pine, scrub, sandhill, and scrubby flatwoods, that occur in patches of 4 ha (10 ac) or more. The model also includes surrounding natural areas and low-intensity land within 2.4 km (1.5 mi) of these habitats if the soils can support xeric ecosystems and related ecological processes. This connectivity helps link two or more patches of xeric natural communities. The 1.5 mi distance was chosen based on dispersal and home range data for key species like the Florida scrub-jay (*Aphelocoma coerulescens*) and gopher tortoise (*Gopherus polyphemus*) (Breininger 1999; Eubanks et al. 2003; Coulon et al. 2010; Hoctor et al. unpublished).
3. The sea-level rise coastal gradient connectivity model identifies compatible land cover within 3.2 km (2 mi) of a projected 3 m sea-level rise, aiming to support inland migration of coastal habitats for ecological resilience. In Florida, protecting functional corridors in coastal zones is challenging due to coastal development, which often isolates wetlands from inland habitats, and all remaining opportunities to protect func-

tional coastal to inland habitat gradients are considered important conservation priorities (see chapter 6).

4. The Integrated Habitat Network (IHN) in southwest Florida has been part of the FEGN since its inception. The IHN aims to preserve and restore riparian connectivity in southwest Florida's phosphate mining region (FDEP 1992). The IHN connectivity model extends 200 m beyond its boundary to include natural, seminatural, agricultural, and extractive lands slated for restoration under postmining regulations.
5. The Florida panther habitat connectivity model uses a panther habitat model (Frakes et al. 2015) to identify the best corridors for Florida panthers, with a minimum width of 1.6 km (1 mi) and some areas exceeding 16 km (10 mi) (Kautz et al. 2006).
6. The Florida black bear habitat connectivity model uses the black bear habitat suitability model developed by the University of Maryland (Poor et al. 2020), which determines optimal corridors with comparable widths to the panther connectivity model for supporting bear movement statewide.
7. The general landscape (structural) connectivity model addresses any remaining network gaps among identified hubs and linkages and seeks to connect all major conservation areas.

In the final step, other areas of ecological significance are connected to the ecological network, including existing conservation lands and Florida Forever proposed conservation lands. Florida Forever proposed conservation lands are approved projects waiting for protection through the Florida Forever process (FNAI 2024b) and represent areas with high natural resource value.

FEGN PRIORITIZATION

The FEGN is also prioritized to guide strategic planning and focus protection efforts on the most irreplaceable and vulnerable areas (see chapters 1 and 3). The initial prioritization, completed in 1999 (Hoctor et al. 2000), has been revised five times, including a major update in 2021 (Hoctor et al. 2025). The prioritization process evaluates critical conservation areas based on ecological significance and connectivity. Key attributes such as habitat quality, species presence, and landscape connectivity are considered to assess each area's role in maintaining ecological functions and supporting biodiversity.

Initially, areas were categorized into high, moderate, or low significance based on factors like the presence of wide-ranging species. Over time, this system evolved into a more detailed seven-tier hierarchy, which was later consolidated into the current five-tier scheme. The current prioritization classes are as follows:

PRIORITY 1 (Critical Linkages)—areas of highest significance for statewide ecological connectivity, including large hubs and most critical corridors. There are 10 critical linkages that identify the most essential ecological connections throughout Florida.

PRIORITY 2—lands supporting Priority 1 areas by expanding functional corridors and enhancing connectivity or providing additional statewide corridors of significance.

PRIORITY 3—major alternative routes to Priority 1 corridors that provide additional options for maintaining ecological links.

PRIORITY 4—important riparian corridors and regional landscapes critical for connectivity.

PRIORITY 5—other significant intact landscapes within the FEGN boundary that meet conservation criteria.

While all FEGN priority levels are vital for maintaining a connected landscape, these priorities guide Florida's conservation land protection programs and land-use planning at both state and local levels. The FEGN focuses on statewide ecological connectivity. Regional to local scale priorities may be addressed in other planning initiatives, and the University of Florida Center for Landscape Conservation Planning is currently working on additional analyses to address ecological connectivity beyond the boundaries of the FEGN to assist regional to local planning.

THE 2021 FEGN

The current version of the FEGN was completed in 2021 (Hoctor et al. 2025), refining previous versions of the network (fig. 7.2; table 7.2), with a new version to be completed in 2025. The FEGN spans 23.1 million ac in total, of which 49% are protected conservation lands in federal, state, local, and private ownership (FNAI 2024a). An additional 10% of the network, amounting to 949,046 ha (>2.3 million ac), is in proposed conservation lands currently pending acquisition through programs such as Florida Forever and the Rural and Family Lands

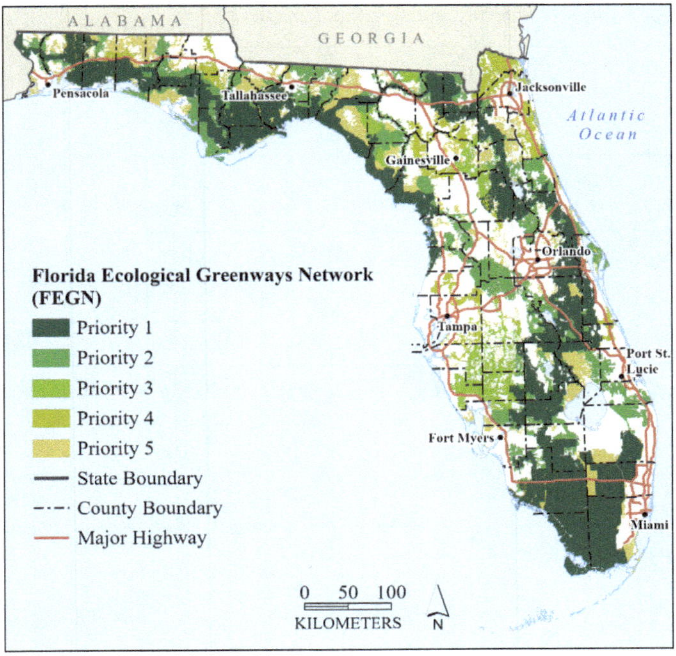

FIGURE 7.2. The updated 2021 FEGN map shows conservation priorities ranked from 1 to 5. Priority 1–3 areas are of the highest ecological significance and comprise the Florida Wildlife Corridor.

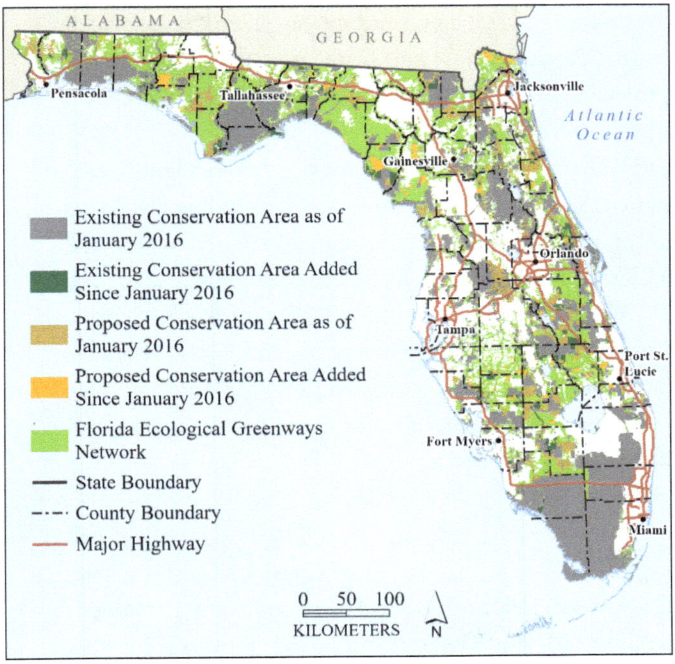

FIGURE 7.3. Additions to the FEGN since 2016, showcasing the proposed parcels for acquisition by the Florida Forever Board of Trustees and the Rural and Family Lands programs.

TABLE 7.2. Land category statistics for the 2021 FEGN boundary

LAND PROTECTION CATEGORY	ACRES	HECTARES	PERCENT	SOURCE
Protected land and water	11,337,344	4,588,060	49.09	CLC 3.7, FLMA 202406
Proposed conservation	2,345,143	949,046	10.15	FFBOT 202406, RFLPP 202311, FLMA 202406
Other private and unprotected land and water	9,414,433	3,809,886	40.76	CLC 3.7, FLMA 202406, DFIRM 100YFP 202404
TOTAL	23,096,920	9,346,992	100.00	

Protection Program. Private, unprotected natural and rural lands comprise 41% of the network covering 3,809,886 ha (9.4 million ac). Despite significant progress in land protection, approximately half of the area within the FEGN boundary remains unprotected, underscoring the continued need for concerted conservation efforts.

The longevity of the FEGN, its continued development through scientific and community partnerships over three decades, and its use in multiple conservation and land-use planning applications have contributed to its acceptance and increasing influence in Florida conservation efforts. The collaborative approach of involving the FEGN TAG ensures that the model reflects the latest scientific understanding and addresses practical conservation needs. The FEGN's iterative modeling process and strategic prioritization exemplify a rigorous and adaptive approach to landscape-scale conservation that effectively identifies and prioritizes a resilient and connected ecological network.

THE FLORIDA GREENWAYS PLAN

The creation of the FEGN was a strategic response to concerns over the fragmentation and degradation of Florida's landscapes. Planning for Florida's greenways and trails began in 1995 with the passage of the initial greenways legislation (*F.S.* 253.787 and 260.012) (FDEP 2024a, b). A five-year implementation plan, completed in 1998 and approved by the legislature in 1999, became the blueprint for implementing a statewide system of greenways and trails (FDEP and FGCC 1998). The implementation plan included the FEGN and five trail networks: hiking, off-road bicycling, equestrian, multiuse trail, and paddling. The Florida Greenways Plan envisions a comprehensive approach to protecting a statewide ecological network while also providing a statewide system of trails that would achieve regional landscape-scale conservation goals for biodiversity and ecosystem services as well as a statewide recreational system. The Florida Greenways Plan, like the FEGN, continues to evolve as trail development and land protection initiatives progress, while some opportunities are lost to development. The Office of Greenways and Trails in the Florida Department of Environmental Protection (FDEP) administers the Florida Greenways Plan, with oversight from the advisory Florida Greenways and Trails Council, while the University of Florida Center for Landscape Conservation Planning continues to update and maintain the FEGN.

THE FEGN IN CONSERVATION AND LAND-USE PLANNING

Florida has long prioritized conserving its natural resources and vital habitats through a comprehensive approach supported by legislative initiatives and funding mechanisms, as detailed in chapter 3. Essential programs such as Florida Forever and the Rural and Family Lands Protection Program (RFLPP) are instrumental for protecting the state's natural ecosystems and rural lands. The FEGN is a primary guide for evaluating and prioritizing land protection projects within these programs. The FEGN has also been an integral part of both state and federal land conservation planning partnerships, including the Florida Fish and Wildlife Conservation Commission's (FWC's) Cooperative Conservation Blueprint from 2008–2014 (FWC, undated), the US Fish and Wildlife Service's Landscape Conservation Cooperatives (USFWS 2021), Natural Resources Conservation Service's (NRCS's) wetland and agricultural land easements (NRCS undated a, b), and the Department of Defense's Sentinel Landscapes (USDA et al. 2024). The FEGN is the foundation for land conservation initiatives with public and private partners to identify specific wildlife corridor projects within the FEGN that are feasible based on landowner interest and available program funding.

Additionally, the FEGN is an essential source of information for CLIP, which identifies conservation priorities for biodiversity, water resources, and functional landscapes (Oetting et al. 2016; see chapter 3). Produced by the University of Florida, FNAI, and FWC, CLIP is regularly updated, and CLIP data are often used in regional to local conservation and land-use planning applications. The Florida legislature continues to renew and increase funding for Florida Forever and the RFLPP, and the FEGN is a primary conservation priority layer used to evaluate and prioritize land protection projects in these programs. The FDOT is also making significant progress planning transportation projects to avoid, minimize, or mitigate future impacts and to ameliorate the impacts of existing roads on ecological connectivity (see chapter 8). In addition, the role of the FEGN in Florida land-use planning has now been further strengthened by the establishment of the Florida Wildlife Corridor, which will be discussed in the following sections.

Outreach to Policy: The Florida Wildlife Corridor and Florida's Adoption of an Official Statewide Wildlife Corridor Plan

THE FLORIDA WILDLIFE CORRIDOR GENESIS

In 2010 a group of ecologists, wildlife biologists, and private lands conservation advocates commissioned a watercolor map by the artist Mike Reagan. The map depicted Priority 1 and 2 (P1 and P2) classes of the FEGN model layered over the state of Florida, with the title "Florida Wildlife Corridor" in the center, next to a small inset portrait of a Florida panther, the state's most iconic wide-ranging carnivore. The commissioning of the map was a first step in a new campaign of the same name, built around the P1 and P2 subsets of the FEGN statewide connectivity model. The idea for the Florida Wildlife Corridor (the Corridor) campaign first emerged at a black bear science workshop in late 2009 hosted at Archbold Biological Station. Drawing inspiration from a presentation of the FEGN Critical Linkages model by Tom Hoctor, conservation photographer Carlton Ward Jr. proposed to the gathering an effort to unify the Critical Linkages into a single concept that would appeal to a broader audience.

Over the next two years the campaign took shape. The effort was broadly informed by conservation science, especially the FEGN Critical Linkages work by the University of Florida's Center for Landscape Conservation Planning. In 2009 a team of black bear researchers from the University of Kentucky, including Joe Guthrie, radio-collared a bear named M34 in Highlands County. The bear's subsequent 805 km (500 mi) dispersal voyage through inhospitable habitats to establish a new home range underscored the necessity of wildlife corridors and a unified conservation strategy for achieving or maintaining connectivity.

The impetus for the Corridor campaign and the new approach emphasizing storytelling was a desire for reinvigorated state funding for land conservation. Prior to 2008 Florida Forever and RFLPP were considered highly successful and well-funded. Florida Forever historically received $300 million annually, supporting land acquisitions and conservation easements critical to conserving and maintaining wildlife and natural habitats. During the economic recession of the late 2000s and the subsequent budget cuts, annual allocations dropped sharply. For several years following the 2008 recession, the Florida Forever program saw scant or no dedicated funding. Similarly, RFLPP, which helps protect working agricultural land, saw inconsistent support. Chronic underfunding by the Florida government weakened the state's ability to protect critical habitats. The organizers of the Corridor campaign saw this downtrend and sought to popularize an ambitious vision for land protection.

THE STATE EXPEDITIONS

After the launch of the Corridor campaign, Ward organized and led a team of conservationists, including Mallory Dimmitt and Joe Guthrie, on a series of statewide and regional expeditions, which gained substantial publicity for the campaign. Complemented by films and additional media, the expeditions engaged the public through storytelling and underscored the Corridor's ecological significance. In 2012, the first expedition team, including filmmaker Elam Stoltzfus, traversed more than 1,600 km (1,000 mi) up the Florida Peninsula from Everglades National Park to Okefenokee Swamp to highlight the ecological connections still remaining. A second 1,600 km expedition in 2015 explored the Florida Big Bend and Panhandle, again using the Critical Linkages layer as a guide to traverse "opportunity" (i.e., unprotected) habitats between large hubs of conservation land, such as Apalachicola National Forest and Eglin Air Force Base in the western Panhandle. Documented by Eric Bendick and Grizzly Creek Films, these expeditions illustrated the fragility, beauty, and conservation significance of Florida's natural and rural landscapes.

REGIONAL EXPEDITIONS

Following the two statewide expeditions designed to highlight the Florida Wildlife Corridor, the campaign's attention turned to regional areas to address distinct at-risk opportunities within the Corridor. These regional expeditions identified critical ecological areas and gaps in the conservation framework, allowing for tailored strategies to maintain the Corridor's integrity. Florida regions covered by these expeditions included the tenuous corridor connecting the Everglades to the Green Swamp through Reedy Creek Swamp, the biodiversity of the Lake Wales Ridge and surrounding ranchlands, the Nature Coast and its rivers and springs north of Tampa, and the O2O corridor, which connects the Ocala National Forest in Central Florida to the Osceola National Forest and Okefenokee National Wildlife Refuge in North Florida and Georgia. These expeditions highlighted both these special landscapes and people interacting with these landscapes while discussing obstacles and opportunities for protection. The regional expeditions and their associated films have informed conservation efforts by discussing specific ecological needs and celebrating local cultural heritage and are planned to continue as part of the work of the Florida Wildlife Corridor Foundation and partners.

FILMS AND PUBLIC MEDIA CAMPAIGN

Ward, Dimmitt, and Guthrie collaborated to promote the Corridor's vision by sharing their documentary films with government officials and decision-makers. Their efforts and the film *Saving the Florida Wildlife Corridor* were pivotal educational tools to encourage the passage of the Florida Wildlife Corridor Act in 2021. Since 2016, Ward has focused on the *Path of the Panther* project, supported by grants from the National Geographic Society and other partners. This project, which used professional photography and video camera traps, aimed to boost support for the Corridor by focusing on the endangered Florida panther. Ward's eight-year effort culminated in the documentary *Path of the Panther*, which attracted national and international attention to the Corridor's conservation needs and sparked widespread public engagement. At the time of this writing, the film has won an Emmy for Outstanding Nature Documentary.

PASSAGE OF THE FLORIDA WILDLIFE CORRIDOR ACT

The Florida Wildlife Corridor team met with legislators and the governor to emphasize the economic, ecological, and social benefits of protecting these landscapes. Their outreach included presentations, meetings, and advocacy campaigns to build legislative support and shape policy. These efforts led to remarkable and bipartisan support for the Corridor, resulting in unanimous passage of the Florida Wildlife Corridor Act in 2021. The Florida Wildlife Corridor campaign was the culmination of decades of dedicated work by scientists, conservation organizations, agencies, landowners, politicians, and engaged stakeholders. The Florida Wildlife Corridor acknowledges the FEGN as its science foundation and is defined by the top three FEGN priorities (P1–P3). The act officially recognized the Corridor's critical role in Florida's economy, cultural and natural heritage, protection of imperiled species, and provision of ecosystem services including clean water. Most importantly, the act reenergized legislative support for land conservation funding. Regular and ample funding is necessary to support Florida Forever, Rural and Family Lands Protection, and Florida Communities Trust, all of which play essential roles in protecting the Corridor and the FEGN.

FEGN BIODIVERSITY SIGNIFICANCE

The FEGN (including the top three priorities within the Corridor) is a crucial tool for protecting Florida's native biodiversity, as habitat loss and fragmentation are the primary factors driving species decline. The FEGN aims to protect rural landscapes across large areas that support viable focal species populations and promote genetic interchange and diversity. It also includes large-scale opportunities to protect coastal to inland habitat gradients in a state known for coastal development and threatened by sea-level rise.

The primary FEGN building blocks, PEAs, focus on high-biodiversity priority areas, landscape-scale species habitat, and natural communities that support viable populations and ecological processes including fire and flood regimes. Identification of PEAs is based on Landscape Species habitat models from FNAI, FWC, and UF for the Florida panther, Florida black bear, crested caracara (*Caracara plancus*), Florida sandhill crane (*Antigone canadensis pratensis*), short-tailed hawk (*Buteo brachyurus*), swallow-tailed kite (*Elanoides forficatus*), southern fox squirrel (*Sciurus niger niger*), Big Cypress fox squirrel (*Sciurus niger avicennia*), eastern indigo snake (*Drymarchon couperi*), eastern diamondback rattlesnake

(*Crotalus adamanteus*), wood stork (*Mycteria americana*), snail kite (*Rostrhamus sociabilis*), Gulf sturgeon (*Acipenser oxyrinchus desotoi*), and Florida manatee (*Trichechus manatus latirostris*).

The matrix-landscape natural communities PEA layer identifies all areas of matrix upland natural communities, wetlands, and large blocks of any natural community type. The upland matrix communities include sandhill, pine flatwoods, dry prairie, upland hardwood forest, upland pine, and slope forest. Wetlands—such as swale, slough marsh, basin marsh, strand swamp, marl prairie, basin swamp, and hydric hammock—are identified in patches of approximately 200 ha (500 ac) or larger as landscape natural community PEAs. Additionally, all natural community blocks of any type 400 ha (1,000 ac) or larger are identified. This combination incorporates nearly all the large remaining blocks of natural habitat in Florida into the FEGN.

Other biodiversity PEA criteria include FWC's Strategic Habitat Conservation Areas, FNAI's Rare Species Habitat Conservation Priorities, FNAI's Underrepresented Natural Communities, FNAI's Potential Natural Areas, and UF's Air Force Conservation Partnership Priority Focal Species Habitat. These layers represent much of Florida's biodiversity conservation priorities, which are then augmented by the FEGN connectivity analyses, including xeric habitat, major riparian corridors, and coastal to inland connectivity, to identify additional landscape gradients that may support species movement. Biodiversity and other resource priority areas not encompassed by the FEGN, generally because they are too isolated, are covered by CLIP (see chapter 3). Nevertheless, the FEGN contains 89% of the high and medium-priority biodiversity areas in the CLIP Biodiversity Resource Category.

FEGN ECOSYSTEM SERVICE SIGNIFICANCE

The Florida Ecological Greenways Network (FEGN) provides vital ecosystem services that benefit human populations. Its key provisioning and regulating services are supplying freshwater by protecting watersheds and aquifers, which supports water supply, fisheries, and tourism. It also provides regulating services by providing flood storage through natural floodplains and wetlands that absorb excess water, improve water quality, mitigate climate change by storing carbon, and moderate local temperatures and humidity. In addition, coastal natural communities and adjacent rural lands within the FEGN offer storm protection by buffering against coastal storm impacts. One of the FEGN layers, Coastal Barrier

Resources Act lands, helps minimize infrastructure damage from coastal storms and is protected under the Coastal Barrier Resources Act of 1982.

The FEGN PEAs also prioritize conservation of essential water sources, including high-priority wetlands, floodplains, and water body buffers. The Riverine/Riparian Corridor Buffers and Connectivity analysis highlights key areas around major rivers where 800 m (2,625 ft) buffers enhance ecological connectivity and protect water resources and habitats. The PEA Natural Floodplains layer identifies areas crucial for flood management and riparian biodiversity, while the PEA Functional Wetlands layer identifies the highest-priority wetlands for habitat provision, water storage and filtration, functional flooding processes, and carbon sequestration. Other ecosystem services, including food and fiber provision and resource-based recreation, are covered in the next section.

NETWORK ECONOMIC / SOCIAL VALUES

Agriculture

The FEGN incorporates many of Florida's agricultural lands, focusing on low-intensity working lands, including silviculture and ranching (table 7.2). Ranching or grazing land constitutes 42.3% of the FEGN's agricultural area and 14.3% of the total FEGN area. Additionally, the network includes 1,589,158 ha (3.9 million ac) of silviculture, representing 50.2% of its agricultural land and 17.0% of the overall FEGN area.

These working lands are crucial for maintaining remnant habitats and connected landscapes essential for wildlife and provide ecosystem services, including flood storage, water filtration, and carbon sequestration as well as food and fiber provision. Voluntary easements and other conservation measures within the FEGN aim to ensure that agricultural lands are protected and managed in ways that align with the network's conservation goals. Studies confirm that improved and semi-improved (seminatural) pastures actively grazed by cattle serve as preferred habitat for several iconic and imperiled grassland birds in Florida, including the Florida sandhill crane, Florida burrowing owl (*Athene cunicularia floridana*), and crested caracara (Morrison and Humphrey 2001), hypothetically because these birds coevolved with megaherbivores in the Pleistocene or earlier and are able to forage more easily in short-cropped grassland (Noss 2012). Silvicultural lands provide forested habitats often embedded with natural wetlands that also often have significant wildlife and ecosystem service conservation values.

Resource-Based Recreation

In 2022, Florida's outdoor recreation economy contributed 3.6% of the state's GDP, surpassing the US average of 2.2% (BEA 2023). The Florida Wildlife Corridor is home to 75 state parks and 32 state forests, harbors over 10,000 km (6,300 mi) of trails, including 1,770 km (1,100 mi) of the Florida National Scenic Trail, and 2,736 km (1,700 mi) of paddling routes, for such activities as hiking, horseback riding, and kayaking.

Key landmarks, including International Dark Sky Parks (Kissimmee Prairie Preserve State Park and Big Cypress National Preserve), offer unique opportunities for stargazing and night sky conservation, and the FEGN model specifically includes dark sky regions as one of the PEA criteria. Additionally, nationally recognized sites such as Canaveral National Seashore, Fort Matanzas National Monument, Everglades National Park, and Big Cypress National Preserve, all located within the Corridor, provide vital spaces for ecotourism, wildlife observation, and nature-based education.

Florida state parks and trails generate over $3.6 billion in revenue and support over 50,000 jobs (FDEP 2023). Expanding access to currently unprotected lands for sustainable outdoor activities, such as camping and other experiences, could further boost conservation and local economies (Florida Wildlife Corridor Foundation, undated).

Clean Water = Tourism (Freshwater, Coastal, Beaches)

Florida's waters—beaches, springs, wetlands, lakes, rivers, and estuaries—play a crucial role in its reputation as a top tourist destination by supporting iconic species and diverse recreational activities. The health of these aquatic ecosystems underpins key tourism activities like boating, paddling, snorkeling, fishing, and nature watching, making healthy ecosystems essential for Florida's tourism economy.

The Corridor includes 55% of Florida's springs, significant parts of 23 springsheds, and over 1,300 rivers and streams, providing critical habitats for species like manatees, Gulf sturgeon, and many other listed aquatic species. Its rivers flow into estuaries that nurture fish populations and provide valuable recreation. Currently, 77% of estuarine areas and 62% of rivers are conserved, with many protected within the Corridor. By connecting and safeguarding these ecosystems, the Corridor supports clean water, biodiversity, and climate resilience, ensuring ongoing tourism and economic benefits for the state (Florida Wildlife Corridor Foundation, undated; Graham et al. 2022).

Implementation of the FEGN / Florida Wildlife Corridor: What Will It Take and Where Are We Headed?

Protecting the FEGN and the Florida Wildlife Corridor requires a multifaceted approach. Central to this effort is collaboration with willing landowners to promote conservation-compatible land management. Building broad support through outreach and education ensures that citizens, stakeholders, and policymakers all understand the important contribution private lands make to long-term biodiversity and ecosystem service conservation.

LAND CONSERVATION FUNDING

Approximately 3.2 million ha (8 million ac) in the Florida Wildlife Corridor remain unprotected; an additional 1.6 million ha (4 million ac) of unprotected land are in the lower-priority FEGN levels 4 and 5. With Florida's population growing by 300,000 annually, over 24,000 ha (60,000 ac) of rural land lost each year to development, and a projected rate of approximately 28,000 ha (70,000 ac) lost through 2070 (Hoctor et al. 2023), there is an urgent need to protect the most critical areas.

Florida's conservationists are engaged in a race to protect at least the most strategic, critical, and functional parts of the Corridor and FEGN before they are lost to development. Securing sustainable funding is paramount for land acquisition and management. Both public and private investments are essential, with programs like Florida Forever and the RFLPP at the forefront. Recently, the RFLPP has received more robust funding to protect agricultural easements (which allow agriculture but protect land from development). We recommend a minimum of $250 million a year for each program to ensure that the state is spending a core minimum of $500 million annually on our two primary statewide conservation land protection programs. The need for this level of funding is accentuated by the fact that there are already around 910,000 ha (2,252,000 ac) of projects in Florida Forever and 150,000 ha (371,000 ac) of projects in the RFLPP. Numerous well-vetted conservation projects are awaiting funding, with many willing landowners ready to participate.

Recent legislative efforts have allocated substantial state resources to conservation, including the Florida Wildlife Corridor Act (2021) (Brodeur 2021), with an average of over $500 million a year for conservation and agricultural land protection from 2021 to 2023. In addition, SB 1638 (2024) (Hutson 2024) directed gaming revenues from the Seminole Tribe Indian Gaming Revenue Trust Fund toward environmental resource management, predominantly to fund conservation land acquisitions and easements within the Corridor.

Local funding initiatives further enhance state and federal efforts. Referenda in 28 out of the state's 67 counties have secured resources for protecting environmentally sensitive lands, parks, open spaces, and working lands during the period 1996–2024, with these programs often attracting matching state and federal funds (Chapin 2011). The Florida Wildlife Corridor legislation has also encouraged these counties to renew existing land protection programs or initiate new ones (Brodeur 2021; Hutson 2024).

Federal programs make a critical contribution, providing significant funding to protect conservation priority areas throughout Florida. The Wetland Reserve Easement program, managed by the Natural Resources Conservation Service (NRCS), supports wetland restoration and protection, providing habitat, water storage, and carbon sequestration benefits. Additional NRCS programs, such as the Agricultural Land Easement Program, protect agricultural lands and significant grasslands, including south-central and southwest Florida's working ranches and prairies in the Everglades and Charlotte Harbor (a National Estuary) watersheds. The NRCS Regional Conservation Partnership Program, working with a variety of land trusts and other partners, has also had a large impact on rural and agricultural land conservation in priority areas across Florida (see chapter 9). The US Forest Service contributes through forest conservation partnerships, including the Forest Legacy Program, using purchases and easements with willing landowners.

INCENTIVE STRATEGIES AND PROGRAMS

Incentives for protection and management primarily include payments for ecosystem services and habitat management. These payments, typically structured like leases for specific periods, do not guarantee permanent land protection but offer economic benefits to rural landowners, especially those with small profit margins in agriculture or natural resources. For example, ecosystem service markets currently operating in Florida compensate qualifying private landowners for protecting species habitat and green infrastructure that stores and cleans water. Such programs can encourage landowners to build relationships with government agencies and conservation organizations, potentially leading to permanent conservation agreements (e.g., easements). Additionally, federal tax benefits for donating land or easements provide economic incentives for permanent protection. Although not feasible for all landowners, tax incentives can be valuable and should be maximized to complement Florida's conservation efforts, with land trusts serving an essential role in promoting such agreements (see chapter 9).

GREEN INFRASTRUCTURE-BASED PLANNING

A common belief is that land protection conflicts with local economic benefits. However, science and economics show that protecting green infrastructure is essential for sustaining healthy, economically vibrant communities (Benedict and McMahon 2006). Green infrastructure—which includes water treatment, flood storage, storm protection, clean air, and parks—provides ecosystem services (Shi et al. 2021) essential for healthy, sustainable human communities. Green infrastructure planning fosters an approach that supports functional ecosystem services, climate resilience, and greenspace for compatible recreation and recognizes the importance of natural capital for healthy, viable communities (Chenoweth et al. 2018). A key challenge is encouraging local governments to adopt planning strategies that protect green infrastructure along with development design that minimizes impacts and facilitates sustainability. Providing economic and tax data showing that smarter, denser development patterns are often better for local economies (including sustainable tax revenue) and the environment is essential (Urban3 2024).

PARTNERSHIPS BETWEEN CONSERVATION AND AGRICULTURE

Of the Opportunity Areas (currently unprotected lands) within the Corridor and in FEGN P4 and P5 outside the Corridor, 80% are in forestry or ranching. In addition, the FEGN includes 68% of the ranchland and 87% of the silvicultural land in the state. Protecting these areas depends on strong partnerships between state and federal agencies, conservation organizations, agricultural landowners, and agricultural commodity associations. Collaboration between conservation and agriculture is crucial for meeting Florida's conservation goals and supporting a sustainable agricultural economy, as well as for ensuring food security. Efforts to enhance agriculture-conservation compatibility include managing ecosystem services and adopting precision technologies, including AI-driven moisture and nutrient monitoring and virtual fencing. Watershed and wildlife habitat management are important cobenefits of well-managed agriculture, and conservation groups, agricultural operators, and agencies are committed to increasing compatibility and sustainability of future Florida agriculture that will also protect the FEGN and Corridor.

BOTTLENECK AND FOCAL AREA PLANNING

Much of the Corridor and FEGN can be protected through "traditional" land protection programs like Florida Forever and RFLPP. However, areas with fragmented ownership, scattered development, narrow corridors, or high development pressure require immediate attention. Such bottlenecks or focal areas are a priority for protection efforts. The University of Florida, the Florida Wildlife Corridor Foundation, Florida Conservation Group, Live Wildly, and other partners are working on local and regional planning strategies to address these challenges. Engaging landowners and stakeholders is key to developing community-supported solutions. Currently, planning or financial incentives for innovative planning in such focal areas are lacking. Legislative action is needed to create incentives for community-driven planning to complement existing conservation funding.

THINKING BEYOND STATE BOUNDARIES

The need for ecological networks extends beyond Florida's borders into Georgia, Alabama, and beyond. River systems that flow from these states into Florida's Gulf Coast highlight the interconnectedness of these regions. Additionally, priority longleaf pine landscapes span across state lines into Georgia and Alabama, including the Red Hills north of Tallahassee, the Trail Ridge that extends from the eastern edge of the Okefenokee Swamp into northeastern Florida, and the Blackwater-Conecuh conservation landscape shared with southern Alabama (FNAI and UF CLCP 2023). Encompassing these regions and extending northward to eastern Canada is the Eastern Wildway mapped by Wildlands Network (see chapter 10).

The University of Florida, Florida Natural Areas Inventory, and other partners are collaborating with universities, state agencies, and conservation organizations in Georgia and Alabama. They are working on GIS analyses and planning strategies to identify priority areas for ecological connectivity across state boundaries and determine the best ways to protect them. These efforts are crucial for aquatic ecosystem protection (including many listed species), restoring longleaf pine landscapes, and facilitating climate adaptation by providing functional connectivity for species-range shifts.

ONLINE PLANNING TOOLS

In the past decade, many geospatial mapping platforms have moved to cloud-based formats, enhancing access to FEGN data. Following the 2021 Florida Wildlife Corridor Act, the UF CLCP and GeoPlan Center developed the FEGN map viewer and dashboard to support land planners, legislators, landowners, and other stakeholders. The map viewer includes data layers for political boundaries, conservation lands, development scenarios, and sea-level rise projections. It allows users to toggle between datasets and perform fundamental analyses. The dashboard highlights protection opportunities and threats, offering visual tools and statistics on development trends and protections related to state, county, and other jurisdictional areas. The FEGN map viewer is available online (UF CLCP 2024a) as well as the FEGN dashboard (UF CLCP 2024b). The UF CLCP and GeoPlan Center are also collaborating on online viewers and dashboards to provide additional data on Florida ecological connectivity priorities and other conservation priority areas, as well as more information on development projections, agricultural lands, and ecological connectivity across state lines.

Conclusion: The FEGN and Florida Wildlife Corridor in Their Science and Policy Context

The FEGN and Florida Wildlife Corridor are central to the protection of Florida's natural landscapes and biodiversity. These initiatives, influenced by scientific research and state policies, aim to integrate ecological knowledge with conservation strategies amid rapid population growth and habitat loss. Florida was the first state to adopt principles of conservation biology and landscape ecology to guide a statewide plan for conserving focal species and ecosystem services using GIS-based modeling. The FEGN is grounded in a science-based approach involving ecologists, conservation biologists, and land management experts. Successful conservation requires integrating top-down and ground-up approaches, with top-down strategies involving broad planning and policy frameworks and ground-up approaches emphasizing local knowledge and community engagement. Strong public-private partnerships are essential for effective conservation, and working lands, such as ranching and forestry, are vital for regional landscape-scale conservation. The FEGN and the Florida Wildlife Corridor demonstrate the importance of a collaborative, science-driven approach to conservation and provide an example of how properly implemented science can serve as a foundation for regional and state-scale conservation planning.

Literature Cited

Bureau of Economic Analysis, US (BEA). 2023. Outdoor Recreation Satellite Account. https://apps.bea.gov/scb/issues/2023/02-february/0223-outdoor-recreation.htm. Accessed October 1, 2024.

Benedict, M. A., and E. McMahon. 2006. Green Infrastructure: Linking Landscapes and Communities. Island Press, Washington, DC.

Breininger, D. R. 1999. Florida Scrub-Jay demography and dispersal in a fragmented landscape. Auk 116:520–527.

Brodeur, J. 2021. Florida Wildlife Corridor Act (CS/CS/SB 976): Protection of ecological systems. Environment and Natural Resources Committee (EN), Florida Senate.

Chenoweth, J., A. R. Anderson, P. Kumar, W. F. Hunt, S. J. Chimbwandira, and T. L. Moore. 2018. The interrelationship of green infrastructure and natural capital. Land Use Policy 75:137–144. https://doi.org/10.1016/j.landusepol.2018.03.021.

Coulon, A., J. W. Fitzpatrick, R. Bowman, and I. J. Lovette. 2010. Effects of habitat fragmentation on effective dispersal of Florida Scrub-Jays. Conservation Biology 24:1080–1088.

Cox, J., R. Kautz, M. MacLaughlin, and T. Gilbert. 1994. Closing the gaps in Florida's wildlife habitat conservation system: Recommendations to meet minimum conservation goals for declining wildlife species and rare plant and animal communities. Florida Game and Fresh Water Fish Commission, Tallahassee.

Eubanks, J. O., W. K. Michener, and C. Guyer. 2003. Patterns of movement and burrow use in a population of gopher tortoises (*Gopherus polyphemus*). Herpetologica 59:311–321.

Florida Department of Environmental Protection (FDEP). 1992. Integrated Habitat Network. Florida Bureau of Mine Reclamation. https://floridadep.gov/sites/default/files/IHN_RegionalConceptualPlan_So_Phosphate_District_FL.pdf. Accessed October 1, 2024.

Florida Department of Environmental Protection (FDEP). 2023. 2022–2023 Economic

Impact Assessment Report Florida State Park System. https://floridadep.gov/sites/default/files/2023%20EIA%20FINAL%20REPORT.pdf. Accessed October 1, 2024.
Florida Department of Environmental Protection (FDEP). 2024a. Office of Greenways and Trails. https://floridadep.gov/ogt. Accessed October 1, 2024.
Florida Department of Environmental Protection (FDEP). 2024b. Historical Timeline of the Florida Greenways and Trails System. https://floridadep.gov/sites/default/files/FGTS%20Historical%20Timeline_2.pdf. Accessed October 1, 2024.
Florida Department of Environmental Protection (FDEP), and Florida Greenways Coordinating Council (FGCC). 1998. Connecting Florida's Communities with Greenways and Trails. Office of Greenways and Trails, Florida Department of Environmental Protection. https://floridadep.gov/sites/default/files/1998FGTSPlanConnectingFlorida%27sCommunities_0.pdf. Accessed October 1, 2023.
Florida Fish and Wildlife Conservation Commission (FWC). Undated. Florida's Cooperative Conservation Blueprint. https://myfwc.com/conservation/special-initiatives/blueprint/. Accessed October 1, 2024.
Florida Natural Areas Inventory (FNAI). 2023. Florida Forever Conservation Needs Assessment. https://www.fnai.org/conslands/ffcna-main. Accessed October 1, 2024.
Florida Natural Areas Inventory (FNAI). 2024a. Conservation Lands. https://www.fnai.org/conslands/conservation-main. Accessed October 1, 2024.
Florida Natural Areas Inventory (FNAI). 2024b. Florida Forever BOT projects. https://www.fnai.org/shapefiles/ffbot_meta_202406.htm. Accessed October 1, 2024.
Florida Natural Areas Inventory (FNAI) and University of Florida Center for Landscape Conservation Planning (UF CLCP). 2023. Longleaf Pine Sustainability Analysis Version 1 Technical Report. Tallahassee, Florida. https://www.fnai.org/PDFs/LSA_v1_Technical_Report_Nov_2023.pdf. Accessed March 25, 2025.
Florida Resources and Environmental Analysis Center (FREAC). 1990. Florida Rivers Assessment: The fifty major rivers of Florida. Florida Resources and Environmental Analysis Center, Florida State University, Tallahassee.
Florida Wildlife Corridor Foundation. Undated. Connecting the Corridor. https://floridawildlifecorridor.org/about/about-the-corridor/. Accessed October 1, 2024.
Frakes, R. A., R. C. Belden, B. E. Wood, and F. E. James. 2015. Landscape analysis of adult Florida panther habitat. PLOS One 10:e0133044. https://doi.org/10.1371/journal.pone.0133044. Accessed October 1, 2024.
Graham, W. D., K. J. Schlatter, A. E. Braswell, M. Brenner, M. J. Cohen, M. J. Deitch, T. G. Gebremicael, A. B. Shortelle, and M. C. Sukop. 2022. Florida Wildlife Corridor Water Benefits Report. University of Florida Water Institute, Gainesville.
Hoctor, T. S. 2003. Regional landscape analysis and reserve design to conserve Florida's biodiversity. Dissertation. University of Florida, Gainesville.
Hoctor, T. S., E. Bohnett, M. O'Brien, R. Noss, S. Lockhart, M. Volk, P. Carr, D. J.

Smith, J. M. Guthrie, and J. Oetting. 2025. Refining the Florida Ecological Greenways Network for improved landscape planning and conservation prioritization. Landscape Ecology 40:36. https://doi.org/10.1007/s10980-025-02049-5.

Hoctor, T. S., M. H. Carr, and P. D. Zwick. 2000. Identifying a linked reserve system using a regional landscape approach: The Florida Ecological Network. Conservation Biology 14:984–1000.

Hoctor, T., P. Carr, P. Zwick, and D. S. Maehr. 2004. Identification and realization of a Florida Ecological Network: The Florida Statewide Greenways Project and its political context. Pages 222–250 in R. H. G. Jongman and G. Pungetti, editors. Ecological Networks and Greenways: Concept, Design, Implementation. Cambridge University Press, Cambridge, UK.

Hoctor, T., M. H. Teisinger, P. Carr, and P. D. Zwick. 2001. Ecological Greenways Network Prioritization for the State of Florida. Office of Greenways and Trails. Florida Department of Environmental Protection, Tallahassee.

Hoctor, T., M. Volk, D. Farrah, and M. O'Brien. 2023. Sea Level 2040/2070. Florida's Rising Seas: Mapping Our Future. University of Florida Center for Landscape Conservation Planning and 1000 Friends of Florida, Gainesville.

Hutson, T. 2024. CS/SB 1638: Funding for Environmental Resource Management. https://flsenate.gov/Session/Bill/2024/1638. Accessed October 1, 2024.

Kautz, R., R. Kawula, T. Hoctor, J. Comiskey, D. Jansen, D. Jennings, J. Kasbohm, F. Mazzotti, R. McBride, L. Richardson, and K. Root. 2006. How much is enough? Landscape-scale conservation for the Florida panther. Biological Conservation 130:118–133.

Lodge, T. 2016. The Everglades Handbook: Understanding the Ecosystem, 4th edn. CRC Press, Boca Raton, Florida.

Morrison, J. L., and S. R. Humphrey. 2001. Conservation value of private lands for Crested Caracaras in Florida. Conservation Biology 15:675–84.

National Resource Conservation Service (NRCS). Undated a. Agricultural Conservation Easement Program. https://www.nrcs.usda.gov/agricultural-conservation-easement-program. Accessed October 1, 2024.

National Resource Conservation Service (NRCS). Undated b. Wetland Reserve Easements. https://www.nrcs.usda.gov/programs-initiatives/wre-wetland-reserve-easements. Accessed October 1, 2024.

Noss, R. F. 2012. Forgotten Grasslands of the South: Natural History and Conservation. Island Press, Washington, DC.

Noss, R. F., and L. D. Harris. 1986. Nodes, networks, and MUM's: Preserving diversity at all scales. Environmental Management 10:299–309.

Oetting, J., T. Hoctor, and M. Volk. 2016. Critical Lands and Waters Identification Project (CLIP): Version 4.0. https://conservation.dcp.ufl.edu/wp-content/uploads/2024/06/CLIP_v4_technical_report.pdf. Accessed October 1, 2024.

Poor, E. E., B. K. Scheick, and M. Mullinax. 2020. Multiscale consensus habitat modeling for landscape level conservation prioritization. Scientific Reports 10:17783. https://doi.org/10.1038/s41598-020-74716-3. Accessed October 1, 2024.

Shi, X., M. Qin, B. Li, and D. Zhang. 2021. A framework for optimizing green infrastructure networks based on landscape connectivity and ecosystem services. Sustainability 13:10053. https://doi.org/10.3390/su131810053. Accessed October 1, 2024.

University of Florida Center for Landscape Conservation Planning (UF CLCP). 2024a. Florida Ecological Greenways Network (FEGN) Map Viewer. Center for Landscape Conservation Planning. https://experience.arcgis.com/experience/27b8d8cd961c4f5091655c52c74d3ada. Accessed October 1, 2024.

University of Florida Center for Landscape Conservation Planning (UF CLCP). 2024b. Florida Ecological Greenways Network (FEGN) Dashboard. University of Florida Center for Landscape Conservation Planning. https://ufl.maps.arcgis.com/apps/dashboards/0caf1487f546449899e2ceda01484a9b. Accessed October 1, 2024.

Urban3 (2024) Economics of Development in Florida. https://1000fof.org/wp-content/uploads/2024/11/EconomicDevelopmentFlorida_Report_v2_compressed.pdf. Accessed March 25, 2025.

US Department of Agriculture (USDA), United States Department of Defense (DoD), and United States Department of Interior (USDI). 2024. Northwest Florida Sentinel Landscapes. https://sentinellandscapes.org/landscapes/northwest-florida/. Accessed October 1, 2024.

US Fish and Wildlife Service (USFWS). 2021. A new landscape conservation task force is established between the U.S. Fish and Wildlife Service and the Association of Fish and Wildlife Agencies. https://www.fws.gov/press-release/2021-12/new-landscape-conservation-task-force-established-between-us-fish-wildlife. Accessed October 1, 2024.

8

Road Ecology Planning and Practice

Daniel J. Smith

Introduction

Roads are recognized as a major cause of habitat loss and landscape fragmentation, posing a threat to connectivity and to core natural areas critical to the maintenance of biodiversity (Noss and Cooperrider 1994; Forman et al. 2003; Watts et al. 2007; van der Ree et al. 2011, 2015). Roadways act as a movement barrier or filter to many species (Forman and Alexander 1998), reducing the extent of accessible habitat and the quality and quantity of available habitat (Eigenbrod 2007).

Roads increase animal mortality from collisions with vehicles (Smith 1996) and lead to avoidance behavior in many species (Forman and Alexander 1998), including carnivores such as black bear (*Ursus americanus*; Brody and Pelton 1984), jaguar (*Panthera onca*; Colchero et al. 2011), and bobcat (*Lynx rufus*; Lovallo and Anderson 1996) as well as birds (Laurance 2004; Kociolek et al. 2011). Overall, the documented negative impacts of roads on populations of species outweigh the documented positive impacts by at least five times (Fahrig and Rytwinski 2009).

Other road impacts include disturbance to adjacent wildlife habitat from roadside lighting, traffic lights, and noise (Reijnen et al. 1997; Francis et al. 2009; Grace and Noss 2018) and degradation of adjacent habitats from roadside dust and other airborne particulates and numerous stormwater effects (e.g., soil erosion, runoff and transport of waterborne sediments and chemicals, altered hydrology, and flooding; Forman et al. 2003). Roads act as conduits to transport invasive plants and animals (e.g., fire ants [*Solenopsis invicta*], cogongrass [*Imperata cylindrica*]), which in turn intrude into adjacent habitats and displace native species (Forys et al. 2002; Gelbard and Belnap 2003). Roads provide human access for hunting (legal and illegal), fishing, and recreation, which can negatively impact wildlife populations (Forman and Alexander 1998). The most destructive impact perpetrated through road access, however, is the opportunity for land conversion and development (Forman et al. 2003; Smith 2003), which increases habitat loss, fragmentation, and negative edge effects.

The recognition of rapid expansion of national and global road networks in the last half of the twentieth century and their potential widespread effects on biodiversity conservation by conservation biologists and landscape ecologists in the 1980s and 1990s generated new ideas and theories about roads and their role in landscape mosaics (Forman 1995). These ideas led to increased efforts to study the ecological impacts of roads, and a new area of research called road ecology emerged.

What Is Road Ecology?

First described by Forman et al. (2003), road ecology is essentially the study of interactions between transportation and natural ecosystems. Road ecologists endeavor to increase understanding of the relationship between the two and to develop and evaluate the effectiveness of planning and mitigation measures designed to avoid or minimize the deleterious effects of roads on wildlife and ecosystems.

Since the early 2000s, road ecology has become a rapidly growing subdiscipline of ecology. This is confirmed by a search of the Web of Science database for key words (ecology, wildlife, habitat fragmentation, edge effects, wildlife connectivity, wildlife corridor, wildlife crossing) combined with road(s) from 2005 to 2024, generating a total of 9,031 references. The same key word searches from 1985 to 2004 produced 882 references. The tenfold increase indicates a significant recognition of the importance of this line of research. Of the many elements of road ecology, this chapter primarily focuses on research associated with the effects of roads on landscape connectivity, wildlife mortality and movement patterns, and applications in the state of Florida.

FIGURE 8.1. A wildlife technician documenting a white-tailed deer involved in a vehicle collision.

Research Methods to Identify Wildlife-Road Conflicts

Many factors influence where wildlife-road conflicts might occur. The most obvious and direct means of locating wildlife-road conflict locations is by collecting animal-vehicle collision (AVC) data. Typically referred to as "counting roadkill" (fig. 8.1), it is a common monitoring technique (Andrews et al. 2015; van der Ree et al. 2015). From collection of these data over time, road ecologists create spatial models of roadkill hotspots (van der Ree 2015; Shilling and Waetjen 2015).

While AVC data and roadkill hotspots models can tell us where road-related animal mortality occurs most often, this is only part of the picture. Concentrations of roadkill may represent areas where many individuals are also crossing successfully, or alternately, they may represent only unsuccessful crossings. It is important to know successful and unsuccessful wildlife crossing locations so that proposed improvements to existing roads can be properly planned to maximize permeability for wildlife (Andrews et al. 2015). For example, consider a location of low traffic volume where animals successfully cross a road today; if an increase in traffic is projected in 5–10 years, it may increase difficulty for animals to cross successfully at that same location.

FIGURE 8.2. An example of hair snare wire placement roadside and successful snag (see inset).

To identify successful road crossing locations, other methods are required (Andrews et al. 2015; van der Ree et al. 2015). Examples of passive techniques include (1) roadside track surveys, (2) point or transect observational (visual or audio [in the case of birds or frogs]) surveys at specified distances from the roadway, (3) camera traps, and (4) landscape genetics, using hair snares to identify DNA (fig. 8.2). Active methods for identifying successful road crossing locations include (1) GPS radiotelemetry, (2) capture-mark-recapture, and (3) use of pit-tags and remote, automated detectors.

Utilizing multiple data types and analysis techniques can increase predictability and accuracy of study results (Andrews et al. 2015; van der Ree et al. 2015). One example is the use of three types of data on black bears to identify hotspots of road crossings (successful and unsuccessful) on SR 40 in the Ocala National Forest (fig. 8.3).

It is common practice to analyze the data collected on wildlife interactions with roads using geographic information systems (GIS). Using GIS enables better understanding of the influence of roads and the juxtaposition and configuration of adjacent land-cover types and features on spatial and temporal patterns of wildlife behavior and movement. Specific metrics on roads and traffic play an

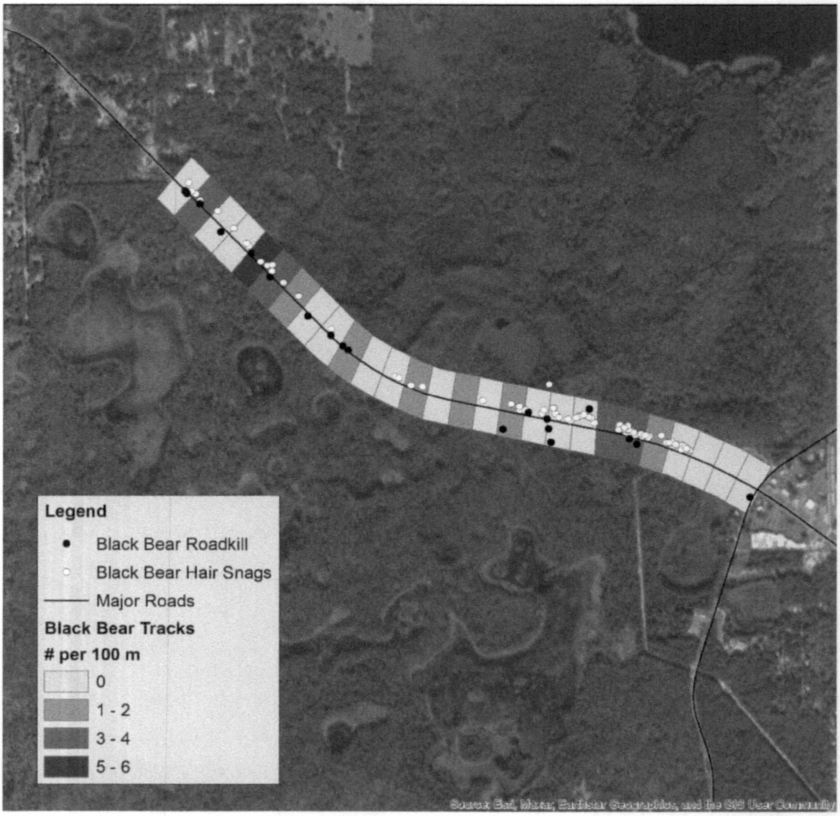

FIGURE 8.3. An example of using roadkill, tracks, and hair snares to identify locations of frequent road crossings by black bears (Smith et al. 2015).

important role in such analyses. Factors include traffic volume, posted speed, number of lanes, road verge width, roadside vegetation, and road geometry (curvature and slope) (fig. 8.4).

Typical GIS analyses used in road impact studies include road crossing behavior, predictive hotspot modeling, home range and habitat use (and how roads may alter these metrics), and wildlife movement modeling (van der Ree et al. 2015). Specific modeling techniques for evaluating landscape connectivity for wildlife movement include resistant kernel, circuitry, least cost path, and linkage mapping (e.g., Bohnett et al. 2024).

Many studies have discussed methods for determining appropriate locations

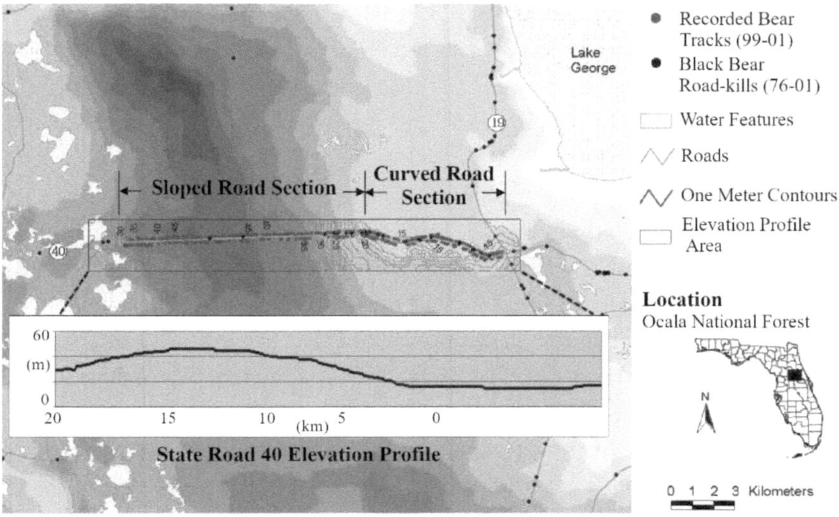

FIGURE 8.4. The relationship between landform, road curvature, and location of AVCs as shown on SR 40 in Ocala National Forest. Tracks represent all crossing attempts, while roadkill represent unsuccessful crossing attempts. Where road curvature and slopes impaired driver visibility, concentrations of AVCs were greatest. Data source: Florida Fish and Wildlife Conservation Commission (FWC); McCown et al. (2004).

for wildlife crossing structures. For example, GIS-based habitat models for species of interest, data on roadkill locations, radiotelemetry, remote camera photos, known migratory paths of animals, and animal signs such as tracks, can identify useful sites for highway crossing structures (Clevenger et al. 2002; Lyren and Crooks 2002; Smith 2012). Siting of crossing structures can benefit from data on unsuccessful crossing locations (i.e., roadkills) but whenever possible should be combined with data on successful crossing locations (i.e., from radio-tracking or tracking stations) and a broader look at the landscape context of the crossing, including the adjacent topography, vegetation, and land use (Smith 2003).

Toolbox for Increasing the Permeability of Roads

Mitigation of road impacts is achieved through the modification of motorist behavior through signage and warnings or the modification of animal behavior through fencing or crossings (Forman et al. 2003; Glista et al. 2009). Unfortunately, standard wildlife signage, speed restrictions, and specialty warning devices have proven relatively ineffective over time (Huijser et al. 2008).

Traffic calming measures such as speed bumps, humps, tables, and traffic roundabouts can be effective in reducing speed considerably and presumably would reduce probabilities for AVCs. While these measures may be impractical and are rarely employed on most high-speed, high-traffic-volume roads, they could be a useful solution on roads in parks and other protected areas. Other experimental measures, such as roadside animal detection/warning systems and integrated smart-road and smart-vehicle networks, may prove effective in time. A few installations of animal detection systems that employ radar and thermal cameras have recently proven effective for large-mammal road crossings (Grace et al. 2015; Huijser, et al. 2017; Gagnon et al. 2018).

Currently, the most proven measures for successfully reducing AVCs and increasing permeability for wildlife are wildlife crossing structures (Clevenger and Huijser 2011; van der Ree et al. 2015; Huijser et al. 2016). The effectiveness of wildlife crossing structures is dependent upon location, crossing design, adjacent suitable habitat, and the use of wildlife fencing (Glista et al. 2009).

Numerous types of crossing structures exist, varying in design, shape, size, and construction material (Clevenger and Huijser 2011; van der Ree et al. 2015; fig. 8.5). The two primary structural types are bridges and culverts with variations in each category. Often described as underpasses, overpasses, and culverts, factors to consider before selecting the type of crossing structure include the target species and their documented preferences and the physical characteristics of the location (van der Ree et al. 2015). To further elaborate, species of vertebrates differ in their requirements and behavioral preferences, such that a given crossing will be permeable to some species but not to others (Glista et al. 2009), potentially causing changes in predator-prey relationships and other community- or ecosystem-level properties (Clevenger and Waltho 2000). Different animals variably react to factors such as openness, adjacent vegetation, light, substrate, noise levels, and presence of humans near the crossings (Glista et al. 2009; Tigas et al. 2002). Several studies have shown that small and medium-sized mammals and many species of amphibians and reptiles use concrete culverts and drainage tunnels (Smith 1996, 2003; Ng et al. 2004).

There are several site design considerations when planning the installation of a wildlife crossing, including (1) context-sensitive design (design with nature concepts; e.g., use native vegetation consistent with surrounding habitat and minimize slope severity of underpass/overpass approaches to improve wildlife

FIGURE 8.5. An array of images of different wildlife crossing structure designs employed in dry and wet environments.

visibility), (2) substrate (soils) consistent with adjacent habitat, (3) adaptation to environmental variation (to provide for terrestrial passage at semiaquatic sites and take into account changing hydroperiod and flood regime), (4) vegetative screening (block headlights, diminish traffic noise, and provide cover), (5) use of berms (reduce effects of traffic noise and lights), (6) fencing (essential to funnel wildlife through passage and away from road surface), (7) lighting (reduce tunnel effects by increasing openness value [height × width / length] and providing light penetration in medians of divided highways), and (8) human presence (manage human access to reduce potential disturbance at crossing sites) (Friedman 1997; Andrews et al. 2015; van der Ree et al. 2015).

Functional design and landscape considerations associated with spacing of wildlife crossings should focus on entire ecological gradients. As the habitat gradient changes from lowland to upland, presence, movement, and dispersal direction of various species may differ (Harris and Gallagher 1988); therefore, we should not only bridge natural barriers to roads such as rivers or wetlands.

Existing bridges and culverts not designed specifically for wildlife movement may nevertheless be used by wildlife, especially when suitable habitat for the species in question exists on either side of the highway (Smith 2003; Ng et al. 2004; Andrews et al. 2015). However, poorly designed structures, such as small-diameter or flooded culverts and bridges, are not used by some animals (Beier 1993; Smith 2003; Andrews et al. 2015). The function of existing structures as wildlife crossings must be evaluated site by site. Yet this is a worthwhile endeavor given the hundreds of existing structures that are present in key locations where landscape connectivity is a priority. Assessments of existing structures represent opportunities for already-in-place wildlife crossings, with many only requiring installation of wildlife fencing or landscaping measures to improve their function for wildlife passage (Andrews et al. 2015).

History of Road Ecology Research and Practice in Florida

EARLY OBSERVATIONS AND PRACTICES (BEFORE 1980)

The earliest published accounts of road impacts on wildlife in Florida were observations by herpetologists who regularly used roadkill to obtain snake specimens for research and collections (Carr 1937; Auffenberg 1956; Kauffeld 1957). The records were used to document species locations and relative concentrations and to create geographic range maps. To this point little concern existed over the broad ecological effects of roads, instead narrowly focusing on AVCs. With the development of concepts like ecological systems and the environmentalism movement in the late 1960s, perspectives on roads would change.

Roads and road networks as threats to ecosystem health, population persistence, and species survival was an emerging issue in the late 1960s and 1970s not only in academic circles but also as part of the public environmental movement and enacting of environmental protection laws. In some cases, road impacts were relevant to recovery plans for rare and imperiled species (Lopez et al. 2003; USFWS 2008 [first iteration approved in 1981]), but the development of practical measures to address the fragmenting effects of roads did not begin until the 1980s (Evink 1990).

Over this 20-year period, actions were triggered by the plight of the endangered Florida panther (*Puma concolor*) and the Florida black bear (*Ursus amer-*

icanus floridanus), two large, wide-ranging species that played major roles in advancing our understanding of broad-scale ecological effects of roads in Florida. Another example, though less well known, concerned ecosystem-level issues rather than single species—specifically, the bisection by US 441 of Paynes Prairie State Preserve in north-central Florida and the killing zone this road represented for thousands of reptiles and amphibians (Smith 1996; Dodd et al. 2004). The following section focuses on these three stories and how they prompted other opportunities for success in the twenty-first century.

THE FIRST ECOLOGICALLY BASED SAFE CROSSING MEASURES AND ROAD IMPACT STUDIES (1980–1999)

With a panther population estimated by the US Fish and Wildlife Service (USFWS) as low as 10 individuals in the early 1970s, monitoring the causes of mortality was essential (Nowak and McBride 1974). The panther served as the first example in Florida to demonstrate the value of long-term monitoring of AVCs and using those data, in part, to plan mitigation measures on roads. In the 1980s, one project proposal would transform SR 84 into an extension of I-75 (called Alligator Alley) to create a high-speed East to West Coast connection to assist with hurricane preparedness and evacuation (Logan and Evink 1985; LoBuono 1988). This major highway would potentially exacerbate fragmentation of core habitat and increase the number of panther-vehicle collisions, further straining a population struggling to sustain numbers of around 30 adults (McBride 1985). A radiotelemetry study was conducted, which along with road mortality and land-cover data identified locations along the proposed highway where conflicts were likely after construction. Plans to address these impacts included the construction of 36 wildlife crossings, 23 specifically to maintain connectivity for terrestrial wildlife and the other 13 oriented more toward wetland-dependent and aquatic species (LoBuono 1988; Evink 1990).

Foster and Humphrey (1995) performed postconstruction monitoring in 1990–1992 and recorded use by panthers, bobcats, deer, raccoons (*Procyon lotor*), bears, and alligators (*Alligator mississippiensis*) in addition to other species (e.g., wading birds and humans). Subsequent monitoring confirmed sustained use by a wide variety of species (Land and Lotz 1996; Smith 2013–2015, unpublished data). The success of this project led to construction of several other crossings on other state and local roads in South Florida targeting use by panthers and aiding with species recovery.

The Florida black bear population, like panthers, also experienced significant declines in the 1960s and 1970s. This trend warranted placing the black bear on the state-threatened species list in 1974 (FWC undated a). This designation prohibited hunting or killing of bears and the development of a management plan with objectives and measures designed to increase subpopulation densities. The FWC began recording bear-vehicle collisions in 1978.

By the early 1990s it was recognized that vehicle collisions were a significant source of bear mortality statewide (Wooding and Brady 1987; Gilbert 1996). To address the issue, an experimental wildlife crossing was designed by the Florida Department of Transportation (FDOT) (Evink 1993). The prototype was a 2.4 m × 7.3 m (8 ft × 24 ft) precast, concrete box culvert created as a lower-cost alternative to the wildlife bridges constructed on I-75 for Florida panthers. Three such structures were installed in 1994 and monitored to evaluate performance, two on SR 29 in Florida panther range and one on SR 46 near Wekiwa Springs State Park targeting black bears. Monitoring studies revealed acceptance of all structures by a wide variety of species, including panthers and bears (Land and Lotz 1996; Roof and Wooding 1996). Subsequent studies demonstrated further acclimation and generally higher levels of use (Smith and Noss 2011; Smith 2013–2015, unpublished data). The success of these early wildlife crossing structure projects kick-started future efforts to plan and construct additional crossing structures to reduce wildlife mortality on roads.

Another project that encouraged more widespread efforts to improve landscape connectivity and reduce AVCs in Florida concerns a large, wet prairie ecosystem where significant numbers of snake roadkills were recorded in the mid-1970s and again in the early 1990s (Smith 1996). Sampling surveys on US Highway 441 across Paynes Prairie in Alachua County revealed approximately 2,750 snakes and many other reptiles, amphibians, and small mammals were being killed on the road each year. Visual observations suggested the ratio of successful to unsuccessful road crossings was about 1 in 17. Results from the 1996 study encouraged state park officials to consult with FDOT and discuss corrective measures to reduce AVCs on US 441. These discussions led to the construction of a barrier wall and underpass system (consisting of four existing box culverts and four new, round-pipe culverts). Pre- and postconstruction surveys documented significant mortality, especially for amphibians and reptiles prior to construction of the wall and significant reductions in roadkill postconstruction, except for treefrogs (Dodd et al. 2004).

These early projects also led to changes in FDOT policy in 1993, establishing evaluation for wildlife crossings as part of all state road projects (Watts 1993). Another benchmark was the creation of a GIS-based environmental screening tool for road project planning created in response to the 1999 Federal Transportation Equity Act for the Twenty-First Century (UF Geoplan Center 2022).

Beginning in 1996, FDOT funded new research to take a broader, statewide approach to identifying wildlife-road conflict locations and wildlife crossing structure needs. Smith (2003) assessed potential interfaces between major roads and priority conservation areas for future mitigation (e.g., lengthening existing bridges and enlarging culverts, constructing new wildlife underpasses) within the Florida Ecological Greenways Network (FEGN). Based on 11 criteria identified by a survey of wildlife and transportation experts, the analysis used GIS to model and prioritize locations where greenways and roadways intersect. The highest priority areas identified for mitigation were regionally and nationally significant conservation areas and important riparian corridors. Smith (2003) later performed site assessments of over 1,200 locations identified in the analysis, including existing bridges and culverts functioning as wildlife crossings.

PROGRESS IN THE TWENTY-FIRST CENTURY

An early follow-up to the statewide analysis on highway-greenway interfaces and existing infrastructure assessments was a two-year monitoring study conducted for FDOT to determine the capacity of existing highway drainage-structure designs to function as wildlife passages (Smith 2003). Wildlife use was monitored at 290 field sites across north-central and Panhandle Florida in various landscape contexts. A logistic regression of 20 structural, environmental, and ecological factors affecting use or avoidance of existing crossing structures was used to establish minimum thresholds for crossing designs. The practical use of this study was never realized and should be reevaluated using new data from more advanced monitoring techniques.

The two most notable, regional-scale examinations of landscape connectivity in relation to roads occurred in FDOT Districts One and Five. District One, located in southwest Florida, contains most of the current Florida panther range, while District Five in east-central Florida encompasses the largest subpopulation of Florida black bears. In District One, construction of wildlife crossings and related studies has been driven mostly by Florida panther recovery efforts. Since 2000, 12 wildlife crossings have been constructed and one roadside animal detection system installed on state and local roads in District One. Most

FIGURE 8.6. Location of panther-vehicle collisions on Florida roads (1998 to 2022). Over this 25-year span, the number of roadkill has risen by an average of 27% annually. Data source: FWC 2023.

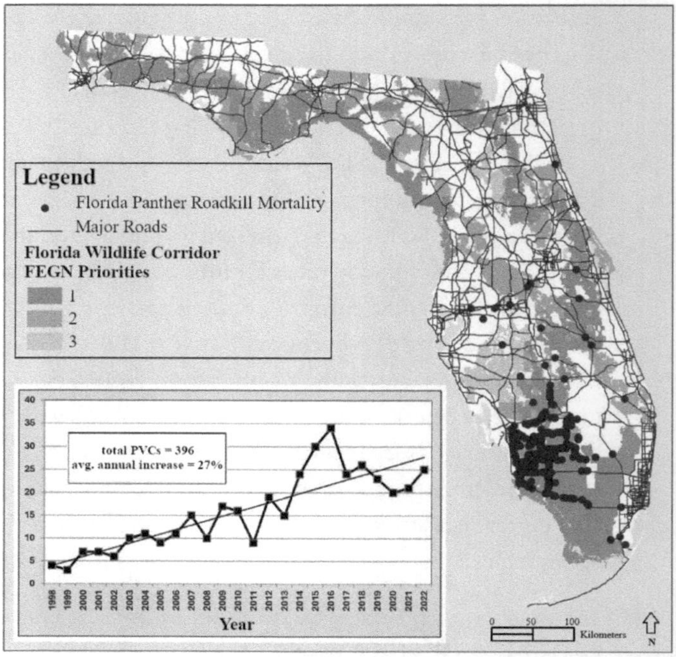

FIGURE 8.7. Location of bear-vehicle collisions on Florida roads (1998 to 2022). From 2013 to 2022, the number of roadkill has averaged 250 annually. Data source: FWC 2023.

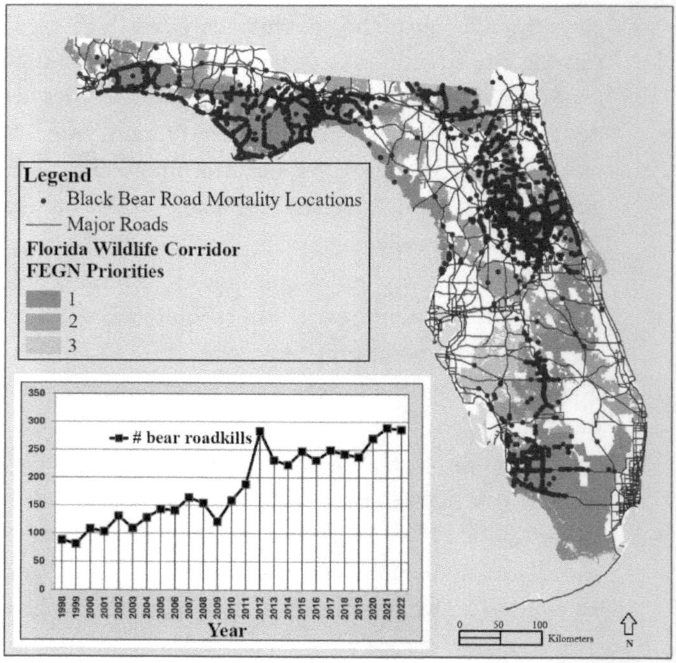

of these included postconstruction monitoring studies (e.g., Smith et al. 2006). There are 18 other wildlife crossings planned within the core habitat range and expansion zone.

The District Five regional wildlife crossing study involved the construction of 13 wildlife crossing structures on six different roads, all located within the Florida Wildlife Corridor. The study included a short period of preconstruction data collection, instead focusing mostly on postconstruction monitoring (Smith and Noss 2011).

Many other site-specific road ecology studies and mitigation projects (including wildlife crossings) have occurred since 2000. Several were dictated by a need to keep pace with continued growth in Florida and to interject functional landscape connectivity into an ever-expanding network of roadways designed to accommodate ever-increasing traffic. Though the Florida panther and black bear populations have rebounded since 2000, increased landscape fragmentation and habitat loss from development and roads have led to more frequent road crossings and dramatic increases in AVCs (figs. 8.6 and 8.7).

These trends are even more illuminating when considering the area requirements for wide-ranging species like panthers and bears. Adult male panther and black bear home ranges are estimated at 194–517 km^2 (USFWS 2023) and 130–310 km^2 (FWC undated b), respectively. Few areas of this size remain in Florida to support populations of individuals requiring such large areas, which are not impeded by roads (fig. 8.8). This exposes individuals to higher risk for involvement in AVCs. Further, the remaining large habitat blocks are considerably disjunct, requiring individuals to traverse a variably dense road network to sustain viable subpopulation densities and maintain genetic interchange between the subpopulations.

As of March 2023, there were 223 existing wildlife crossing structures, four under construction, and 56 planned across the state (fig. 8.9). They range in size from small culverts to long bridges. Although most wildlife crossings built to date addressed panther or bear road mortality, the crossing structures also serve as safe passage for many other species. Several crossing structures have been constructed in association with (1) the state's bridge inventory replacement program (restoring riparian habitat connectivity), (2) reconnecting previously severed wetland ecosystems, and (3) new road construction.

Most notable are the recent major projects associated with limited access, high-speed transportation routes—Interstate 4 (I-4), Wekiva Parkway, and Suncoast Parkway. The design or redesign associated with these roadways was (as much

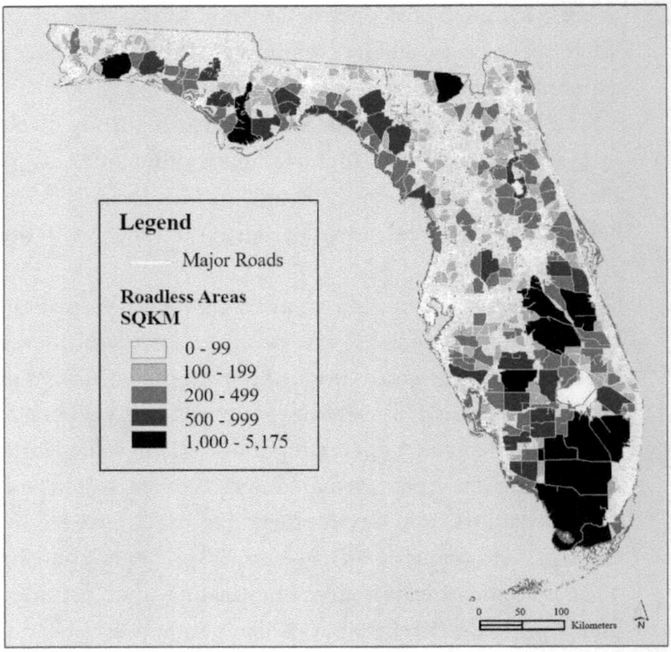

FIGURE 8.8. Roadless areas in Florida. Note that the analysis only includes major roads; local neighborhood streets, and most unpaved roads were excluded. Data source: FDOT 2023.

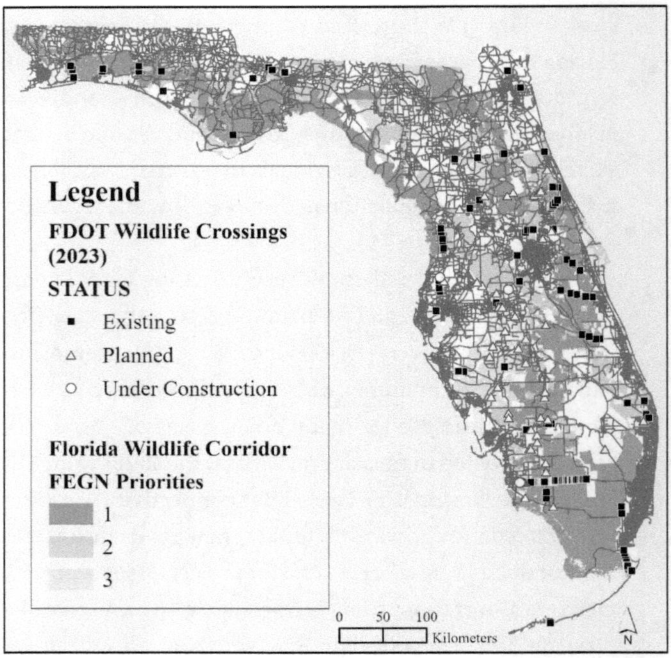

FIGURE 8.9. Existing and proposed wildlife crossing structures in Florida. Data source: FDOT.

as practical) to retain landscape connectivity that existed prior to construction (Wekiva Parkway, Suncoast Parkway) or to restore landscape connectivity as part of reconstruction (I-4). The Wekiva Parkway features four bridges ranging from 244 m (800 ft) to 1,219 m (4,000 ft) long. The Suncoast Parkway includes a mix of 13 wetland and upland wildlife crossing bridges. The most recent wildlife crossing on the Suncoast included pre- and postconstruction monitoring and evaluation. I-4 was reconstructed where it crosses Tiger Bay State Forest and three 30.5 m (100 ft) long large animal crossings and five small animal crossings were installed to allow safe movement for bear and many other species.

The study on Paynes Prairie previously discussed served as an early model for examining multispecies and ecosystem-level road impacts and mitigation planning. A similar project was completed at Lake Jackson in Leon County in 2010, where numerous aquatic turtles were being killed on US 27, which interfered with seasonal migrations between hibernation and breeding sites (Aresco 2005).

Where the Paynes Prairie study fell short was in examining the wildlife-road interactions along the upland fringe of the wetland basin. A subsequent project at Ross Prairie in Marion County (Smith 2003) included these transitional edges (known to be high wildlife activity areas) as well as adjacent uplands. Findings led to the recommendation for two bridges at the wetland-upland ecotones, four upland culverts for gopher tortoise, and construction of a herpetofauna wall and replacement of the single, central cross-drain in the wetland basin with five culverts spaced throughout the wetland to more broadly add connectivity at different hydroperiod levels. Plans also included an equestrian bridge and a recreational trail crossing.

Integrating Transportation and Conservation Planning

LESSONS LEARNED

The projects presented in the previous section serve as a body of work to evaluate current practices and identify opportunities to make improvements and increase efficacy. One important area that is undervalued is monitoring and maintenance of wildlife crossing structures. Too frequently, pre- and postconstruction monitoring is brief and not comprehensive enough to advise and optimize decision-making in choosing locations for wildlife crossings, reviewing and selecting among site and structural design alternatives, and evaluating performance upon completion. Prior to construction, at least two to three years of monitoring should be a standard to account for interannual variability in environmental conditions (e.g., wet and dry years) that may influence presence, movements, and activity of various species (Andrews et al. 2015; van der Ree et al. 2015). Typically, at least five years of monitoring after construction is necessary to allow for acclimation of animals to the structure (Clevenger 2005). An approach prescribed in instances where random assignment of treatment and control sites cannot be applied is a before-after-control-impact/treatment (BACI) design (Andrews et al. 2015; Rytwinski et al. 2015; Schmidt et al. 2021). These are more long-term studies, but the information obtained is valuable for adaptive management and in understanding functionality and then improving designs and cost-effectiveness.

Maintenance should be considered part of adaptive management. It is often overlooked, but essential in subtropical locations like Florida to control invasive and destructive plants and insects, address drainage and erosion problems, inspect and keep fences in shape, and address other issues that may reduce wildlife use and overall effectiveness (Andrews et al. 2015; van der Ree et al. 2015). Typical roadside vegetation management is not applicable at crossings—for example, standard turfgrass mowing cycles. Landscaping at crossings commonly includes an array of native ground covers and shrubs that require a different approach when the goal is to mimic wildlife habitat.

INCLUSION OF CONSERVATION SCIENCE IN TRANSPORTATION PLANNING

Establishing a collaborative environment that brings together experts in the disciplines of ecology, engineering, planning, and conservation is essential to integrating transportation and conservation planning. A communicative, productive, and forward-thinking setting generates more creative and effective solutions to the challenges of establishing and managing an ecologically functional Florida Wildlife Corridor that can be sustained for generations.

In this spirit, several GIS tools have been created to more effectively mesh landscape connectivity and eco-network planning and design with transportation elements. Examples include animal-vehicle hotspots modeling, animal movement models that identify intersections of roads and landscape linkages, and databases that include existing transportation facilities that promote connectivity.

Three such resources include a statewide AVC hotspots model, an inventory of existing bridges that coincide with AVC hotspot locations, and a corridor or least-cost-path model of panther range expansion that identifies potential road

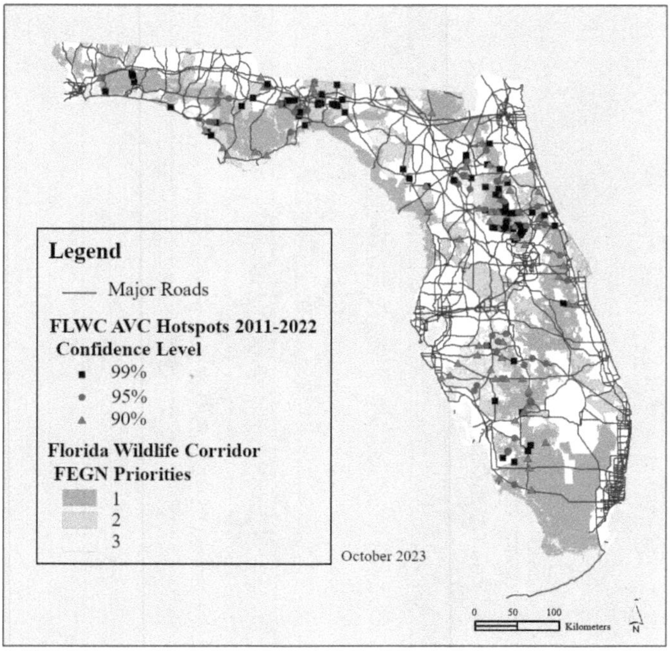

FIGURE 8.10. Large animal-vehicle collisions identified in the Florida Wildlife Corridor. Results include 246 total hotspots within confidence levels 90%, 95% and 99% (Smith and Gagne 2023).

conflict locations. A statewide AVC hotspots model of large animals was created for FDOT in 2021 and later updated to target locations in the Florida Wildlife Corridor in 2023 (Smith et al. 2022; Smith and Gagne 2023). The analysis identified 246 hotspots at three levels of confidence using the Getic Ord Gi* statistic; 104 hotspots were identified at the 99% confidence level (fig. 8.10).

The panther corridor and least-cost-path models were created to (1) predict potential pathways that panthers are likely to use under existing land-cover / land-use conditions from the current species core range to large habitat hubs north of I-4, and (2) identify probable road crossing locations (Smith 2022). We used least-cost-path (LCP) and least-cost-corridor (LCC) functions in ESRI's ArcMap 10.8. The FWC Cooperative Land Cover and USFWS 2012 Panther Habitat Assessment Methodology Classification were used to identify land-cover type preferences for panthers. The original 192 land-cover classes were consolidated into 17 classes closely corresponding to the USFWS classification

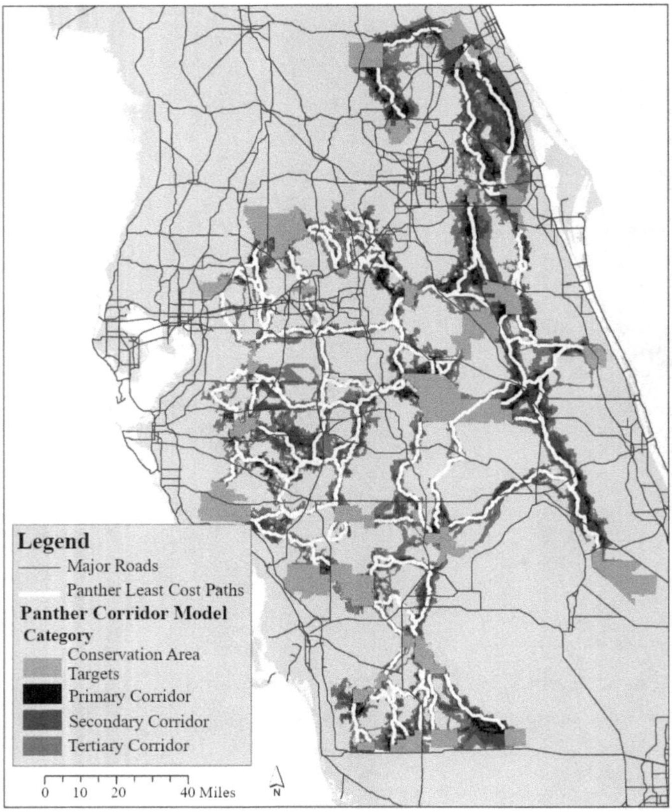

FIGURE 8.11. Florida panther movements and potential road crossing locations within south-central Florida (Smith 2022).

categories. Connections (LCPs and corridors) were modeled between 34 target areas within the FEGN (fig. 8.11). Overall, 86% of the predicted corridors overlap with existing conservation lands or the FEGN. The primary level corridors consist of 94% existing conservation lands or the FEGN. There were 236 road segments (potential conflicts) identified that intersect least-cost paths and 227 bridges (crossing opportunities) within 0.8 km (0.5 mi). These resources were included as data layers in the Environmental Screening Tool that FDOT uses in its Efficient Transportation Decision Making (ETDM) process to identify and address potential environmental impacts of transportation projects (UF Geoplan Center 2022).

FDOT POLICIES AND MEASURES

Collaboration, communication, and action have been at the center of recent efforts by FDOT to work with resources agencies, conservation groups, and the public to establish policies and guidelines that incorporate state conservation objectives on wildlife and wildlife habitat issues and the Florida Wildlife Corridor. These include measures to reduce AVCs, improve landscape connectivity and permeability of our roadways, and facilitate success of the Florida Wildlife Corridor.

FDOT continues to work with partners to update and create new data resources and to provide funding and opportunities for research, planning, and land protection. One useful initiative is establishing a broader, systematic wildlife crossing monitoring program throughout the state that can generate data for research and environmental outreach.

MOVING FORWARD

Roads are one of the greatest threats to wildlife worldwide (van der Ree et al. 2015). Biologists recognize that crossing structures and other measures are needed in many places to allow wildlife to successfully cross highways and maintain connectivity and gene flow within and among populations (Forman et al. 2003; Smith 2003; van der Ree et al. 2015). One impediment is cost. The construction of large-scale crossing structures is expensive, over one million dollars for a single structure (Beckmann et al. 2010; Clevenger and Huijser 2011). Therefore, it is necessary to identify new funding sources but also to refine and expand research and monitoring to accurately determine the number and type of structures needed to reduce costs while maximizing the potential for improving road permeability for wildlife. Importantly, monitoring must encompass multiple species (e.g., amphibians and reptiles as well as mammals) because different structures and landscape/habitat conditions promote movement of different taxa (Clevenger and Waltho 2005; Smith 2003). Planners need to know the locations where wildlife are most likely to traverse roads so that mitigation measures are best employed to reduce collisions and population fragmentation (Langen et al. 2009).

Literature Cited

Andrews, K. M., P. Nanjappa, and S. P. D. Riley. 2015. Roads and Ecological Infrastructure: Concepts and Applications for Small Animals. Johns Hopkins University Press, Baltimore, Maryland.

Aresco, M. J. 2005. Mitigation measures to reduce highway mortality of turtles and other herpetofauna at a north Florida lake. Journal of Wildlife Management 69:549–560.

Auffenberg, W. 1956. A study of the fossil snakes of Florida. PhD Dissertation. University of Florida, Gainesville.

Beckmann, J. P., A. P. Clevenger, M. P. Huijser, and J. A. Hilty. 2010. Safe Passages: Highways, wildlife, and habitat connectivity. Island Press, Washington, DC.

Beier, P. 1993. Determining minimum habitat areas and habitat corridors for cougars. Conservation Biology 7:94–108.

Bohnett, E., J. Oetting, R. Noss, M. O'Brien, R. Frakes, D. Smith, S. Lockhart, J. Mullinax, E. Poor, B. Scheick, and T. Hoctor. 2024. Consolidating diverse modeling methods and spatial prioritization for multispecies connectivity planning. Frontiers in Conservation Science, https://doi.org/10.3389/fcosc.2024.1406944.

Brody, A. J., and M. R. Pelton. 1989. Effects of roads on black bear movements in western North Carolina. Wildlife Society Bulletin 17:5–10.

Carr, A. F., Jr. 1937. The geographic and ecological distribution of the reptiles and amphibians of Florida. PhD Dissertation. University of Florida, Gainesville.

Clevenger, A. P. 2005. Conservation value of wildlife crossings: Measures of performance and research directions. GAIA—Ecological Perspectives for Science and Society 14(2):124–129.

Clevenger, A. P., and M. P Huijser. 2011. Wildlife Crossing Structure Handbook: Design and Evaluation in North America. Report No. FHWA-CFL/TD-11-003. Federal Highway Administration, Washington, DC.

Clevenger, A. P., J. Wierzchowski, B. Chruszcz, and K. Gunson. 2002. GIS-generated, expert-based models for identifying wildlife habitat linkages and planning mitigation passages. Conservation Biology 16:503–514.

Clevenger, A. P., and N. Waltho. 2000. Factors influencing the effectiveness of wildlife underpasses in Banff National Park, Alberta, Canada. Conservation Biology 14:47–56.

Colchero, F., D. A. Conde, C. Manterola, C. Chavez, A. Rivera, and G. Ceballos. 2011. Jaguars on the move: Modeling movement to mitigate fragmentation from road expansion in the Mayan forests. Animal Conservation 14:158–166.

Dodd, C. K., Jr., W. J. Barichovich, and L. L. Smith. 2004. Effectiveness of a barrier wall and culverts in reducing wildlife mortality on a heavily traveled highway in Florida. Biological Conservation 118:619–631.

Eigenbrod, F., S. J. Hecnar, and L. Fahrig. 2008. Accessible habitat: An improved measure of the effects of habitat loss and roads on wildlife populations. Landscape Ecology 23:159–168.

Evink, G. L. 1990. Wildlife crossings of Florida I-75. Transportation Research Record 1279:54–59.

Evink, G. L. 1993. Highways and wildlife. Environews 1(1):3–6.

Fahrig, L., and T. Rytwinski. 2009. Effects of roads on animal abundance: An empirical review and synthesis. Ecology and Society 14(1):21.

Florida Fish and Wildlife Conservation Commission (FWC). Undated a. Black bear research. https://myfwc.com/research/wildlife/terrestrial-mammals/bear/. Accessed September 26, 2024.

Florida Fish and Wildlife Conservation Commission (FWC). Undated b. Facts (about black bears). https://myfwc.com/wildlifehabitats/wildlife/bear/facts/. Accessed September 26, 2024.

Forman, R. T. T. 1995. Land Mosaics: The Ecology of Landscapes and Regions. Cambridge University Press, New York.

Forman, R. T. T., and L. E. Alexander. 1998. Roads and their major ecological effects. Annual Review of Ecology and Systematics 29:207–231.

Forman, R. T. T., D. Sperling, J. A. Bissonette, A. P. Clevenger, C. D. Cutshall, V. H. Dale, L. Fahrig, R. France, C. R. Goldman, K. Heanue, J. A. Jones, F. J. Swanson, T. Turrentine, and T. C. Winter. 2003. Road Ecology: Science and Solutions. Island Press, Washington, DC.

Forys, E. A., C. R. Allen, and D. P. Wojcik. 2002. Influence of the proximity and amount of human development and roads on the occurrence of the red imported fire ant in the lower Florida Keys. Biological Conservation 108:27–33.

Foster, M. L., and S. R. Humphrey. 1995. Use of highway underpasses by Florida panthers and other wildlife. Wildlife Society Bulletin 23:95–100.

Francis, C. D., C. P. Ortega, and A. Cruz. 2009. Noise pollution changes avian communities and species interactions. Current Biology 19:1415–1419.

Friedman, D. S. 1997. Nature as infrastructure: The National Ecological Network and wildlife crossing structures in the Netherlands. Report no. 138. DLO Winand Staring Centre, Wageningen, The Netherlands.

Gagnon, J. W., N. L. Dodd, S. C. Sprague, K. S. Ogren, C. D. Loberger, and R. E. Schweinsburg. 2018. Animal-activated highway crosswalk: Long-term impact on elk-vehicle collisions, vehicle speeds, and motorist braking response. Human Dimensions of Wildlife 24:132–147.

Gelbard, J. L., and J. Belnap. 2003. Roads as conduits for exotic plant invasions in a semiarid landscape. Conservation Biology 17:420–432.

Gilbert, T. 1996. An overview of black bear roadkill in Florida 1976–1995. In G. Evink, P. Garrett, D. Zeigler, and J. Berry, editors. Trends in addressing transportation-related wildlife mortality. Florida Department of Transportation, Tallahassee.

Glista, D. J., T. L. DeVault, and J. A. DeWoody. 2009. A review of mitigation measures for reducing wildlife mortality on roadways. Landscape and Urban Planning 91:1–7.

Grace, M. K., and R. F. Noss. 2018. Evidence for selective avoidance of traffic noise by anuran amphibians. Animal Conservation 21:343–351.

Grace, M. K., D. J. Smith, and R. F. Noss. 2015. Testing alternative designs for a roadside animal detection system using a driving simulator. Nature Conservation 11:61–77.

Harris, L. D., and P. B. Gallagher. 1989. New initiatives for wildlife conservation: The need for movement corridors. Pages 11–24 in G. Mackintosh, editor. In Defense of Wildlife: Preserving Communities and Corridors. Defenders of Wildlife, Washington, DC.

Huijser, M. P., E. R. Fairbank, and F. D. Abra. 2017. The reliability and effectiveness of a radar-based animal detection system. Report No. INE/CESTiCC 17.14. Center for Environmentally Sustainable Transportation in Cold Climates, University of Alaska Fairbanks and US Department of Transportation, Washington, DC.

Huijser, M. P., E. R. Fairbank, W. Camel-Means, J. Graham, V. Watson, P. Basting, and D. Becker. 2016. Effectiveness of short sections of wildlife fencing and crossing structures along highways in reducing wildlife-vehicle collisions and providing safe crossing opportunities for large mammals. Biological Conservation 197:61–68.

Huijser, M. P., P. McGowen, J. Fuller, A. Hardy, A. Kociolek, A. P. Clevenger, D. J. Smith, and R. Ament. 2008. Wildlife-vehicle collision reduction study. Report to Congress. US Department of Transportation, Federal Highway Administration, Washington, DC.

Kauffeld, C. 1957. Snakes and snake hunting. Hanover House, Garden City, NY.

Kociolek, A. V., A. P. Clevenger, C. C. St. Clair, and D. S. Proppe. 2011. Effects of road networks on bird populations. Conservation Biology 25:241–249.

Land, D., and M. Lotz. 1996. Wildlife crossing designs and use by Florida panthers and other wildlife in southwest Florida. In G. Evink, P. Garrett, D. Zeigler, and J. Berry, editors. Trends in addressing transportation-related wildlife mortality. Florida Department of Transportation, Tallahassee.

Langen, T. A., K. M. Ogden, and L. L. Schwarting. 2009. Predicting hot spots of herpetofauna road mortality along highway networks. Journal of Wildlife Management 73:104–114.

Laurance, S. G. W. 2004. Responses of understory rain forest birds to road edges in central Amazonia. Ecological Applications 14:1344–1357.

LoBuono, J. P. 1988. Alligator Alley: Protecting natural habitat. Journal of the Florida Engineering Society 41(2):14–16.

Logan, T., and G. Evink. 1985. Safer travel for the Florida panther. Naturalist 6(3/4):6–7.

Lopez, R. R., M. E. P. Vieira, N. J. Silvy, P. A. Frank, S. W. Whisenant, and D. A. Jones. 2003. Survival, mortality, and life expectancy of Florida Key deer. Journal of Wildlife Management 67:34–45.

Lovallo, M. J., and E. M. Anderson. 1995. Bobcat (*Lynx rufus*) home range size and habitat use in northwest Wisconsin. American Midland Naturalist 135:241–252.
Lyren, L. M., and K. R. Crooks. 2002. Factors influencing the movement, spatial patterns and wildlife underpass use of coyotes and bobcats along State Route 71 in Southern California. Proceedings of the International Conference on Ecology and Transportation. Center for Transportation and the Environment, Raleigh, North Carolina.
McBride, R. T. 1985. Population status of the Florida panther in Everglades National Park and Big Cypress National Preserve. Report to the National Park Service in fulfillment of Contract #RFP 5280-84 04, Homestead, Florida.
McCown, W., P. Kubilis, T. Eason, and B. Scheick. 2004. Black bear movements and habitat use relative to roads in Ocala National Forest. Florida Fish and Wildlife Conservation Commission, Tallahassee, Florida, Final Report. FDOT Contract # BD-016.
Ng, S. J., J. W. Dole, R. M. Sauvajot, S. P. D. Riley, and T. J. Valone. 2004. Use of highway undercrossings by wildlife in southern California. Biological Conservation 115:499–507.
Noss, R. F., and A. Y. Cooperrider. 1994. Saving Nature's Legacy: Protecting and Restoring Biodiversity. Defenders of Wildlife and Island Press, Washington, DC.
Nowak, R. M., and R. T. McBride. 1974. Status survey of the Florida panther. Project 973. World Wildlife Fund Yearbook 1973–74:237–242.
Reijnen, R., R. Foppen, and G. Veenbaas. 1997. Disturbance by traffic of breeding birds: Evaluation of the effect and considerations in planning and managing road corridors. Biodiversity Conservation 6:567–581.
Roof, J., and J. Wooding. 1996. Evaluation of SR 46 wildlife crossing. Technical report no. 54. FLDOT FL-ER-61–96. Florida Cooperative Fish and Wildlife Research Unit, US Biological Service, Gainesville.
Schmidt, G. M., R. L. Lewison, and H. M. Swarts. 2021. Pairing long-term population monitoring and wildlife crossing structure interaction data to evaluate road mitigation effectiveness. Biological Conservation 257:109085.
Shilling, F. M., and D. P. Waetjen. 2015. Wildlife-vehicle collision observation collection and hotspot identification at large scales. Nature Conservation 11:41–60.
Smith, D. J. 1996. The direct and indirect impacts of highways on the vertebrates of Payne's Prairie State Preserve. Technical Report No. FL-ER-62–96. Florida Department of Transportation, Tallahassee.
Smith, D. J. 2003. The ecological effects of roads: Theory, analysis, management, and planning considerations. PhD Dissertation. University of Florida, Gainesville.
Smith, D. J. 2012. Determining location and design of cost-effective wildlife crossing structures along US 64 in North Carolina. Transportation Research Record 2270:31–38.
Smith, D. J. 2022. Identifying Least-Cost Paths and Corridors for Florida panther

within South-Central Florida. Summary Report. Prepared for US Fish and Wildlife Service, Florida Panther Recovery Implementation Team.
Smith, D. J., and C. Gagne. 2023. Large Animal-Vehicle Collision Hotspots on Roads within the Florida Wildlife Corridor. Final Report. University of Florida Center for Landscape Conservation Planning, Gainesville, and Archbold Biological Station, Lake Placid, Florida.
Smith, D. J., C. Gagne, and J. Rote. 2022. Analysis of Animal-Vehicle Collisions on Florida Roads Using the Getis-Ord Gi* Spatial Statistic Tool in ArcGIS. Final Report, UCF-FDOT Project #BDV 24-TWO985-04.
Smith, D. J., M. K. Grace, H. R. Chasez, and M. J. W. Noss. 2015. State Road 40 pre-construction wildlife movement monitoring: Areas A, B and F. Final Report, Contract No. BDK78, TWO #501-3. Florida Department of Transportation, District Five, Deland.
Smith, D. J., and R. F. Noss. 2011. A reconnaissance study of actual and potential wildlife crossing structures in Central Florida, Final Report. Contract No. BDB-10. Florida Department of Transportation, Tallahassee.
Smith, D. J., R. F. Noss, and M. B. Main. 2006. East Collier County wildlife movement study: SR 29, CR 846, and CR 858 wildlife crossing project. University of Central Florida and University of Florida IFAS. Southwest Florida REC Research Report, SWFREC-IMM-2007-01.
Smith, L., and C. K. Dodd Jr. 1999. Determination of the effectiveness of wildlife barriers and underpasses on U.S. Highway 441 across Paynes Prairie State Preserve, Alachua County, Florida. Phase I final report, contract no. BB-854. Florida Department of Transportation, Tallahassee.
Tigas, L. A., D. H. Van Vuren, and R. M. Sauvajot. 2002. Behavioral responses of bobcats and coyotes to habitat fragmentation and corridors in an urban environment. Biological Conservation 108:299–306.
University of Florida (UF) Geoplan Center. 2022. Efficient Transportation Decision Making (ETDM). https://www.geoplan.ufl.edu/portfolio/etdm/. Accessed September 26, 2024.
US Fish and Wildlife Service (USFWS). 2008. Florida Panther Recovery Plan (*Puma concolor coryi*), Third Revision. US Fish and Wildlife Service. Atlanta, Georgia.
US Fish and Wildlife Service (USFWS). 2023. Florida Panther National Wildlife Refuge. Circular No. 508. https://fws.gov/refuge/florida-panther. Accessed September 29, 2024.
van der Ree, R, D. J. Smith, and C. Grilo, editors. 2015. Handbook of Road Ecology. Wiley-Blackwell, Oxford, UK.
van der Ree, R., J. A. G. Jaeger, E. A. van der Grift, and A. P. Clevenger. 2011. Effects of roads and traffic on wildlife populations and landscape function: Road ecology is moving towards larger scales. Ecology and Society 16(1):48.

Watts, B. 1993. Environmental policy for state transportation facilities. EnviroNews 1:1.
Watts, R. D., R. W. Compton, J. H. McCammon, C. L. Rich, S. M. Wright, T. Owens, and D. S. Ouren. 2007. Roadless space of the conterminous United States. Science 316:736–738.
Wooding, J. B., and J. R. Brady. 1987. Black bear roadkills in Florida. Proceedings of the Annual Conference of the Southeastern Association of Fish and Wildlife Agencies 41:438–442.

9

How Land Trusts Implement Landscape Conservation

Susan Carr, Reed F. Noss, Sarah Lockhart, Ramesh Buch, Shane Wellendorf, Debi Osborne, Ken Grudens, Lauren Day, Kimberly Tillman, and Tom Kay

Introduction

As nonprofit organizations dedicated to conservation missions, land trusts use a variety of methods to protect and manage land for conservation purposes as well as provide services such as public education, outreach, and advocacy. Here we describe the conservation tools land trusts use to protect land, especially conservation easements. We review the historic and current influence of land trusts on land conservation with an emphasis on Florida. We focus primarily on conservation planning at the scale of large, heterogeneous ecosystems (landscapes) rather than for individual sites, species, or resources.

A defining characteristic of land trusts is that they own land or interests in land. However, land trusts also accelerate land protection by guiding implementation of public land conservation programs. This is particularly apparent in states and counties that have robust public land conservation programs, such as Florida. We summarize some of the traditional methods that involve acquiring land interests but also discuss how land trusts work with public programs and initiatives, which to some extent distinguishes Florida land trusts from those elsewhere in the nation.

More than 1,700 land trusts operate in the United States today, ranging from small local nonprofits to large national and international organizations such as the Trust for Public Land and The Nature Conservancy (Bray 2010; Cheever and McLaughlin 2015). Most land trusts work in specific regions ranging from a single county to an entire state. Although most land trusts' missions center on land conservation, priorities vary among them. Some land trusts work to conserve specific land types or land uses (such as agricultural lands or traditional land uses) or elements (such as imperiled species), whereas others have a broader focus.

Traditional methods employed by land trusts include acquiring land outright (i.e., fee simple acquisition) and acquiring conservation easements on properties that remain in private ownership. A conservation easement is a voluntary agreement that permanently protects conservation values of a property by restricting development and land conversion. It is a legal agreement that requires both a willing landowner (the grantor) and a holder (the grantee), which is typically a land trust, government agency, or other qualified entity. The grantee is obligated to monitor adherence to easement terms and to enforce them if they are violated. Importantly, land trusts work with perpetual conservation easements—these "run with the land," meaning that both current and future landowners are bound by the easement restrictions. Easements can be either purchased by a land trust or public agency or donated to a qualified grantee by a landowner. Landowners who sell conservation easements receive cash, whereas donated easements may have tax advantages for the grantor.

The term "conservation easement" was coined in the 1950s by journalist William Whyte, who advocated controls on private land use to accomplish conservation objectives (Cheever and McLaughlin 2015). By that time, however, what we now call conservation easements already had been implemented in several parts of the United States. The first such easements were established in the late 1880s to protect parkways designed in the Boston area by Frederick Law Olmsted (Byers and Ponte 2005; Bray 2010). During the 1930s and 1940s, the National Park Service encumbered easements on some 1,500 acres of private land along the scenic route of the Blue Ridge Parkway in Virginia and North Carolina, and another 4,500 acres along the Natchez Trace Parkway in Mississippi, Alabama, and Tennessee (Cheever and McLaughlin 2015).

In 1964, the IRS authorized a federal income tax deduction to landowners who donated conservation easements adjacent to federal highways. One year later the IRS extended tax benefits to landowners who donated easements to government agencies or charitable organizations, such as land trusts. In 1980,

Congress made the conservation easement deduction a permanent part of the tax code (Cheever and McLaughlin 2015).

The use of easements as conservation tools expanded greatly during the 1980s, accompanied by a tremendous growth in the number of land trusts nationwide. Between 1988 and 2003, the acreage protected nationally by land trusts through conservation easements grew by 1,624% (Byers and Ponte 2005). By 2003, some 70% of all land protected annually by The Nature Conservancy in the US was via conservation easements (Fishburn et al. 2009). Today, conservation easements are one of the most popular land protection tools in the United States, and the primary land protection method employed by land trusts. More than 61 million acres have been protected by conservation easements nationally (Land Trust Alliance undated).

Land Trusts and Conservation Planning in Florida

Florida land trusts operate similarly to land trusts elsewhere in the country, with some notable distinctions. First, Florida land trusts are heavily involved with the development and implementation of state and local conservation programs, whereas many other states lack such programs. The early onset of Florida's state conservation programs influenced land trust development, favoring larger organizations that facilitated state acquisitions rather than smaller local community land trusts. More recently, land trust growth is fueled by burgeoning conservation programs at county and city levels as well as increases in federal funding for agricultural land protection. Second, Florida faces the highest rate of human population growth and associated development in the country (see chapters 1 and 4). This heightens the urgency for land protection and inspires political will and public funding, which in turn fuel growth of regional land trusts.

Land trusts continue to play a big role in the development of landscape-scale conservation. As of 2025, 28 land trusts operate in Florida, including 22 that are Florida based and six national organizations (fig. 9.1). Some national and state organizations include land conservation as part of their broader missions, including Ducks Unlimited, Wildlands Conservation, and the Florida Conservation Group. However, most Florida land trusts have operating areas ranging from one to several counties. Smaller land trusts often depend on volunteers, whereas larger land trusts typically employ paid staff. Support for most land trusts comes from individual donors and members, whereas larger organizations also receive

NATIONAL Land Trusts

Name
American Farmland Trust*
Ducks Unlimited*
National Park Trust
The Conservation Fund*
The Nature Conservancy*
Trust for Public Land*

FLORIDA Land Trusts
(at least part of service area in Florida)

Name	Florida Service Area
Alachua Conservation Trust*	North
Apalachee Land Conservancy	Eastern Panhandle
Bay County Conservancy	Bay County
Big Waters Land Trust*	Southwest
Corkscrew Regional Ecosystem Watershed Land & Water Trust	Southwest
Conservation Florida*	Statewide
Calusa Land Trust	Pine Island
Cape Coral Wildlife Trust	Lee County
Florida Conservation Group	Southwest/Central
Georgia-Alabama Land Trust*	North
Green Horizon Land Trust	South Central
Indian River Land Trust	Indian River County
North Florida Land Trust*	Northeast
Orlando Land Trust	Orlando
Putnam Land Conservancy	North Central
Sanibel-Captiva Conservation Foundation	Lee County
Tall Timbers Research Station and Land Conservancy*	Eastern Panhandle
Tampa Bay Conservancy*	Tampa Bay area
The Environmental Conservancy of North Port	Southwest
Treasured Lands Foundation	Martin County
Wildlands Conservation	Central
The Winter Park Land Trust	Winter Park

* Land Trusts accredited by the Land Trust Alliance

FIGURE 9.1. Current list of land trusts operating in Florida. Source: Land Trust Alliance, Alliance of Florida Land Trusts, Sarah Lockhart (unpublished).

support from corporate donations, philanthropic organizations, government contracts, and public grants.

Florida has a robust history of science-based conservation planning (see chapters 2 and 3). Land conservation is largely about opportunity—being at the right place at the right time with the right tools. Systematic conservation planning promotes "informed opportunism, taking advantage of conservation openings as they arise, but with explicit recognition of the trade-offs involved" (Noss et al. 2002). Land trusts work with a variety of plans that guide conservation across multiple purposes and scales. Arguably, Florida's most influential landscape conservation plan is the Florida Ecological Greenways Network (FEGN), with its top three priority layers codified in the Florida Wildlife Corridor Act of 2021 (see chapter 7). Conservation plans guide land trusts' work in many ways, by influencing priority properties and allocation of resources. In addition, conservation plans guide land trust work with landscape restoration, including the America's Longleaf Restoration Initiative and Florida State Wildlife Action Plan.

Many land trusts also focus on more local geographies, including county governments with land conservation programs. An example is the newly established Conservation Lands Acquisition and Management Program of Nassau County (Hudson 2020). This northeast Florida county worked with the North Florida Land Trust to prioritize the protection of their region's natural and cultural resources. Other examples include conservation plans for specific species (e.g., the National Bobwhite Conservation Initiative), plans for agricultural land protection, and municipal plans for recreation and greenspace (see Case Studies section).

As implementors of landscape conservation, Florida land trusts serve many roles: They are land buyers, conservation easement holders, facilitators of public programs, advocates, fund raisers, land stewards, partnership conveners, and technical experts. Furthermore, land trusts help refine landscape conservation strategies, providing feedback and information throughout the dynamic process of conservation implementation.

Brief History of Land Trusts and Landscape Conservation in Florida

Land trusts emerged as significant conservation practitioners in Florida in the 1980s as part of the national land trust movement. This emergence coincided with the creation of state programs that dedicated public funding to the acquisition of conservation lands (Henderson 2022). Population growth in Florida exploded in the mid-twentieth century, triggering rapid development at the expense of natural landscapes. Floridians reacted by pressing their leaders not only for regulatory protections but also for funding to acquire lands. In response, Florida land protection programs grew during the 1970s and 1980s (Henderson 2022). In 1990, the state dedicated $1 billion to fund the "Preservation 2000" land conservation initiative, followed by the Florida Forever program, established in 2000 and funded with another $1 billion (see chapter 3). Importantly, the Florida Forever initiative included direct grants to local and nonprofit partners for land conservation through the Florida Communities Trust (FCT) program, a significant factor in the evolution of Florida land trusts. Furthermore, land trusts were required to meet rigorous acquisition and organizational criteria to be eligible for FCT grants, which strengthened their governance capacity.

Technological advances in the early 2000s prompted rapid development of landscape-scale conservation planning tools, most notably geographic information systems (GIS) and computer-powered data models. Moreover, there were major advances in theoretical and applied conservation science, prompting new approaches and models that underpin conservation planning. These developments brought new ideas, scientific tools, and professionalism to the practice of land conservation.

The Great Recession of late 2007 to mid-2009 brought severe cutbacks to state funding for land conservation. Funding for Florida Forever and other state conservation programs plummeted, and the national land trusts reduced staffing in Florida. A notable exception was the Conservation Fund, which retained staff during this time and executed several large projects. Regional and local Florida land trusts stepped up to fill the void, continuing to work with local government programs and landowners. Public conservation programs have vigorously rebounded in recent years, with near-record-high funding levels for state and federal conservation programs in 2021–2023. Accordingly, land trusts greatly expanded their work with these public programs. Today, land trusts are key players in the implementation of landscape conservation in Florida.

Tools of the Trade: Direct Acquisition of Conservation Lands and Conservation Easements

As noted earlier, land trusts protect land in many ways, including outright purchase (fee simple) and the acquisition of "less-than-fee" conservation interests in properties, usually referred to as conservation easements. Buying land outright can ensure its permanent protection. However, land purchases are costly, tie up funds needed elsewhere, and invoke additional costs in the form of insurance, maintenance, and management. Land purchases by land trusts are generally preceded by fundraising campaigns. This can be an effective way to raise awareness along with the capital needed to buy land. However, there are limits to the number and magnitude of such campaigns. For these (and other) reasons, direct land purchases by land trusts are not as common as other methods of land protection.

Donations of land and conservation easements are important methods of land protection, and these donated land interests are typically a cornerstone of land trusts' portfolios. Other arrangements include bargain sales and life estate transfers, which sometimes involve donations of land interests. As charitable organizations, land trusts accept donated land interests, which, as noted earlier, confer tax advantages to donors. As a method to protect their land while retaining it in family ownership, donated conservation easements appeal to many conservation-minded landowners. Landowners work with land trusts to transact donations and secure the charitable treatment of the gift without the involvement of a public agency. Although donated conservation easements remain an important land protection mechanism, they comprise a relatively small acreage of total conservation lands in Florida. An exception to this generalization is the Red Hills region of the Florida Panhandle, where large landscapes are protected via donated conservation easements (see Case Studies section).

The Land Trust Alliance, a national association, supports land trusts by helping them manage the risks and costs of holding conservation easements. Most Florida land trusts have adopted the Land Trust Alliance Standards and Practices, a comprehensive set of guidelines for ethical and responsible operation (Land Trust Alliance 2024). The Land Trust Accreditation Commission, an independent program of the Land Trust Alliance, administers a program for accreditation of land trusts that pass rigorous evaluation of governance, finances, transactions, and stewardship. To date, six Florida land trusts maintain this accreditation, having met the highest standards for running their organization and conserving and stewarding land. Furthermore, many Florida land trusts participate in Terrafirma, a charitable risk pool collectively owned by land trusts that insures its members against the legal costs of defending conservation easements in the event of violations.

Landowners often are not able to donate land interests. Rather, they seek to sell a conservation easement for their dual goal of receiving payments while protecting their lands from development. Sale proceeds may support land management or be a part of family estate planning. Alternatively, a landowner may want to sell their land outright to become part of a public park. In these cases, the buyer of these lands or conservation easements is a government agency with a well-funded conservation acquisition program.

Land trusts also serve an important role as an intermediary in the acquisition process between willing sellers and a government agency (the conservation buyer). Usually, a landowner wishes to receive fair market value for the sale of their land, and the conservation buyer is competing with other potential buyers, including quick-moving developers and investors. This puts government agencies at a distinct disadvantage because they are bound by eligibility criteria, procedural rules, and transparency requirements. Furthermore, landowner trust in agencies may be lacking. Land trust facilitation of conservation acquisitions benefits both parties. In addition, land trusts are often able to work with landowners before development pressures or market conditions motivate the need for quick sales.

Compared to land purchases, the market differs for conservation easement acquisitions. In essence, the landowner (seller) is monetizing a portion of their land rights and forming a permanent relationship with the conservation easement holder, which extends to all future owners. Conservation easement transactions are fundamentally different from land purchases, with more complexity and lengthy transaction periods. Land trusts play an essential role in guiding landowners through this process.

Public Land Conservation Programs

In this section, we summarize the major public programs that fund conservation land acquisitions in Florida. It is important to note that conservation lands and easements are purchased only from voluntary sellers; no conservation programs in Florida use eminent domain or condemnation to acquire land.

STATE CONSERVATION PROGRAMS

State land conservation programs were reviewed in chapter 3, so we offer only a brief summary here. Florida Forever is the state's premier conservation program and the largest state land acquisition program in the nation. As of 2024, more than one million ha (2.6 million ac) have been protected through Florida Forever and its precursor programs. This represents nearly half of Florida's 2.4 million ha (6 million ac) of state conservation lands (FNAI 2024). Florida Forever is funded from a documentary stamp tax linked to real estate and securities transactions, tying the program's funding to the growth of Florida's real estate market.

The other large state conservation acquisition program is the Rural and Family Lands Protection Program (RFLPP), administered by the Florida Department of Agriculture and Consumer Services (FDACS). This program purchases conservation easements to protect agricultural lands, also known as agricultural easements or agricultural conservation easements. The RFLPP's mission is to sustain agricultural working lands by purchasing permanent easements to preclude development and keep lands in agriculture. After a period of dormancy, the RFLPP was reinvigorated in 2022 with a large legislative appropriation, and by 2024 the program had acquired over 32,000 ha (80,000 ac) of agricultural easements and had over 250 eligible projects on its waiting list (FNAI 2024; FDACS 2024). Given the prevalence of working lands in priority conservation landscapes, the RFLPP is an important contributor to Florida's landscape conservation.

FEDERAL CONSERVATION PROGRAMS

Florida land trusts work with many federal conservation programs, housed in several departments, including the Departments of Defense, Interior, and Agriculture. Here we describe federal programs that provide funding and technical support for land protection and land management. The US Fish and Wildlife Service (USFWS), the agency responsible for managing the nation's fish and wildlife, provides grants for the acquisition and restoration of wetlands through the North American Wetlands Conservation Act, which has supported protection of large wetland complexes in North and Central Florida. The US Forest Service administers the Forest Legacy Program to protect forestland from conversion to nonforest uses. The Land and Water Conservation Fund provides funding to federal agencies to protect land within established priority areas, including the recently designated Everglades to Gulf Conservation Area in South and Central Florida. Land trusts develop multiyear projects around these large federal programs, which often involve multiple partners.

The US Farm Bill is the largest single source of conservation funding in the US and a prominent funding source for conservation and agricultural easements in Florida. The Natural Resource Conservation Service (NRCS) administers Farm

Bill programs that purchase conservation and agricultural easements and provides funding for private land management. NRCS conservation easement programs have various land eligibilities and conservation purposes, but in general they center on the protection of agricultural lands and wetlands. The Agricultural Conservation Easement Program—Wetlands Reserve Easement (ACEP-WRE) program, prominent in Central and South Florida, protects and restores degraded wetlands. ACEP-WRE easements are part of the strategy to restore landscapes in the Kissimmee River basin and the Upper Everglades Headwaters region of south-central Florida.

The 2018 Farm Bill reauthorization prompted greater land trust participation with NRCS conservation easement programs, including the Agricultural Conservation Easement Program—Agricultural Land Easement (ACEP-ALE) and the Regional Conservation Partnership Program (RCPP). These programs provide matching funds for purchases of agricultural and conservation easements held by nonfederal partners, such as land trusts and state and local governments. The ACEP-ALE program is designed to protect productive grasslands such as grazing and ranch lands—land types common in central and southern Florida. In contrast, the RCPP program has broader land eligibility and offers more flexibility but is more complex in its implementation. These programs provide partial funding for purchases of conservation and agricultural easements, matched with nonfederal contributions. Land trusts typically develop and facilitate these transactions. This "blended funding" strategy leads to complicated and lengthy transactions. However, this approach leverages federal resources and is one way to accelerate conservation of Florida's working lands.

In recent years, the Department of Defense (DoD) emerged as a major advocate for, and funder of, land conservation. Florida is home to 42 military facilities, several of which are among the nation's largest and biologically richest. The DoD partners with land trusts to buffer military facilities from incompatible development on adjacent lands as part of the Readiness and Environmental Protection Integration (REPI) program, which funds conservation that benefits the military's training mission. Land trusts work with REPI funding to purchase land and conservation easements adjacent to military bases (see Case Studies section). Similarly, the DoD's Sentinel Landscapes Partnership program encourages public and private coalitions that advance sustainable land-use practices around military installations and ranges.

LOCAL GOVERNMENT PROGRAMS AND INITIATIVES

Florida is exceptional in its number and magnitude of local government programs dedicated to financing land conservation. Since the 1990s, Florida voters have approved 97 local referenda to protect lands for clean drinking water, recreational opportunities, and wildlife habitat. These initiatives created and expanded dozens of local government land conservation programs and generated over $14 billion for land protection. County and city conservation programs bring dedicated and predictable funding and also benefit from expertise and support of local constituents and stakeholders. Voters in Florida coastal counties have repeatedly shown support for land conservation, as they consistently approve county conservation referenda. In more recent years, voters of less-populated counties have similarly approved ballot initiatives.

Land trusts often facilitate the creation of local government conservation programs. For example, the Indian River Land Trust, in collaboration with the Trust for Public Land, led three successful countywide environmental land bond referenda in Indian River County, generating $126 million for protection of native habitat, locally important waterways, and agricultural lands. Two land trusts in North Florida spearheaded conservation programs in three counties: Alachua Conservation Trust (ACT) led to Alachua County's multiple ballot initiatives resulting in over $300 million for the Alachua County Forever program (see Case Studies section). Similarly, North Florida Land Trust worked with Nassau and Clay Counties to bring $75 million for land protection in those counties. These partnerships do not stop after the referenda are passed. Land trusts continue to work with county governments to find willing sellers, facilitate transactions, and leverage other funding sources.

MARKET-BASED CONSERVATION APPROACHES

Market-based conservation tools are another approach to land conservation in Florida. In general, these include economic approaches and monetary incentives to conserve land and resources. These programs involve voluntary exchanges, subject to market forces of supply and demand and economic valuation.

Species conservation banking is a growing method of land conservation in Florida, which is home to many imperiled species (see chapter 1). Increasingly, land trusts work with landowners who participate in conservation banking programs as a method to protect or manage their lands. The Florida Fish and

Wildlife Conservation Commission (FWC) leads the program to protect Florida's population of the state-threatened gopher tortoise (*Gopherus polyphemus*). FWC authorizes gopher tortoise relocations from planned development sites to eligible recipient sites, and landowners are compensated for receiving gopher tortoises at prices determined on the open market. FWC permits often require permanent land protection in the form of a conservation easement. Other similar species mitigation programs (e.g., Florida sand skink [*Plestiodon reynoldsi*] and Florida scrub-jay [*Aphelocoma coerulescens*]) also include long-term land protection. Collectively, these programs offer novel incentives and approaches for private land conservation.

Conservation Easements as the Preeminent Land Protection Tool in Florida

In recent decades, conservation easements have ascended as the primary tool for protecting land in Florida. Nearly all the conservation programs described in the previous sections fund conservation easement purchases. The area protected with conservation easements held by state, federal, and local government agencies is more than 445,000 ha (1.1 million ac), nearly 10% of all of Florida's conservation lands (FNAI 2024). The growing importance of conservation easements warrants discussion of their advantages and disadvantages.

ADVANTAGES OF CONSERVATION EASEMENTS FOR LANDSCAPE CONSERVATION

Conservation easements offer several advantages as an alternative to outright land acquisition. First, easement purchases are less expensive than land purchases, and public funding can therefore buy more acres for conservation. The "more bang for the buck" appeals to taxpayers and policy makers as a way to stretch precious and limited conservation dollars. This also allows conservation programs to prioritize funding for direct purchase of land to protect exceptional habitats or to buy parcels that are critical to the integrity of existing public lands. Another cost-saving advantage is that privately owned conservation lands reduce or eliminate the public cost of land management. Public agencies bear the cost of managing public lands, but land management costs for privately owned conservation lands are typically borne by the landowner. This reduces the financial burden on public land management agencies. Conservation easements can be an effective tool for conserving large swaths of privately owned land in priority landscapes.

The variety of conservation easement programs appeals to a wide array of Florida landowners. Depending on the goals of the program, conservation easements have various degrees of restrictiveness and flexibility. For instance, a restrictive easement may forbid land-use conversion to agriculture, but a less restrictive working lands easement may allow shifts in land use such as conversion from forestry to row crops. This variety in conservation easements can be advantageous as a land protection tool.

Conservation easements allow landowners to benefit financially by selling or donating a portion of their land rights while retaining their ownership. Many conservation easements allow for continued conservation-compatible, income-producing land uses such as timber, agriculture, or grazing. In addition, most Florida conservation easement properties are eligible for state property tax exemptions—another incentive that broadens appeal to landowners. For families interested in maintaining long-term ownership, a conservation easement can be a valuable estate planning tool and assist with the generational transfer of land ownership. Finally, lands protected with conservation easements generally do not require public access. This appeals to landowners with privacy and liability concerns.

Advocates for conservation easements cite the need to balance growth and resource protection, particularly in fast-growing parts of Florida (Archbold Biological Station 2024). For example, conservation easements may be used to protect water resources by deterring urban development without removing farmlands from the land base. These flexibilities, along with the other advantages listed in this section, make conservation easements an attractive method for implementing landscape conservation strategies.

DISADVANTAGES OF CONSERVATION EASEMENTS

The disadvantages of conservation easements as a landscape conservation tool are both real and potential. As a legal instrument, a conservation easement restricts land uses and may motivate good land stewardship (e.g., requiring landowners to control invasive plants). However, purchased conservation easements that restrict land uses do not necessarily obligate landowners to manage their land for conservation outcomes. For example, a purchased conservation easement may restrict the landowner's land-use options (i.e., what they cannot do) but may not compel the landowner to use prescribed fire to improve habitat. "Affirmative easements" require activities on the part of either the easement holder or the landowner, but these are uncommon in purchased conservation easements in Florida.

Requirements for donated easements are generally stricter. Donated conservation easements require that either the donor or the easement holder satisfy specific conservation purposes outlined in the tax code. These include "protection of a relatively natural habitat of fish, wildlife, or plants, or similar ecosystem" or provision for public recreation, scenic enjoyment, or protection of historic areas or structures (McLaughlin 2020). Furthermore, Department of Treasury regulations provide that "relatively natural habitat" must be a "significant habitat," which includes but is not limited to habitats for rare, endangered, or threatened species; high-quality examples of terrestrial or aquatic communities; or natural areas that are included in, or contribute to, the ecological viability of a designated conservation area (McLaughlin 2020). In response to a Senate Finance Committee recommendation in 2005 that "the IRS consider revoking the tax-exempt status of a conservation organization that regularly and continuously fails to monitor and enforce conservation easements," the Land Trust Alliance made the monitoring, defense, and enforcement of easements a key feature of its accreditation process (Halperin 2011).

The most common stated purpose of conservation easements in the United States is open space preservation, which may or may not coincide with protection of biodiversity and other conservation values (Dyckman et al. 2024). Among others, Looney (2017) claims that some donors of easements apply "grossly inflated appraisals to the value of the easement to increase their charitable deduction or by taking donations for easements that do not fulfill bona fide conservation purposes." The controversy over the tax revenue costs of easements has generated increased scrutiny of existing and proposed easements by the IRS (Cheever and McLaughlin 2015). This is a positive development if it ultimately results in properties of higher conservation value being selected for easements and in better monitoring and enforcement of easements. The Land Trust Alliance also requires its member land trusts to refuse to accept donations of easements intended primarily as tax shelters (Lindstrom 2019).

Conservation easements affect current and future property values. Usually, the market value of lands with conservation easements is reduced because of the limitations to land rights. Although a landowner who initially sells or donates a conservation easement enjoys proceeds from the sale, or a tax deduction, conservation easement restrictions may limit opportunities for selling the property for development or receiving income from some land uses; easements therefore lower the land value and limit future sale opportunities. Tax benefits from donations may not fully compensate for this loss of income.

In nearly all cases, conservation easements have withstood legal challenges to erode or extinguish restrictions. Conservation easements are governed by a Florida state statute, which authorizes perpetual easements. Conservation easements held by accredited land trusts, the Florida Department of Environmental Protection, and the Department of Agriculture have provisions that ensure perpetuity and significantly limit the ability to extinguish a conservation easement. In rare circumstances, conservation easements can be extinguished, but only through judicial condemnation and when it is clear that the conservation values of the property can no longer be maintained (McLaughlin 2024). Extinguishment is a complicated and difficult process and is only considered after all other options have been exhausted or a property is irreparably damaged.

How Land Trusts Work with Landscape Conservation Plans and Programs

Successful land conservation depends on the convergence of strategy and opportunity. Out of necessity, Florida land trusts work opportunistically but within a strategic conservation framework. The Florida Wildlife Corridor (the Corridor) represents the best example of this (see chapter 7). The Corridor is composed of protected lands and unprotected lands identified by the Florida Ecological Greenways Network (FEGN). Land trusts whose focal areas lie within the Corridor seek opportunities within the FEGN such as willing sellers, eligible lands, eligible programs, and funding sources.

All Florida land acquisition programs are built on voluntary participation—that is, landowners who are willing to sell or donate land interests or otherwise participate in conservation programs. A parcel may be identified as high priority in a conservation plan, but if the landowner is not a willing participant, not much can be done. In this very common scenario, a land trust may keep tabs on the property to monitor changes in ownership and the owner's outlook while continuing to cultivate the relationship. Conversely, a landowner may be eager to sell, donate, or participate, but their land has low conservation priority—which can also change with time. Land trusts are attuned to such vicissitudes and are positioned to move quickly as opportunities arise.

Land Trusts Provide Information and Data for Refining Conservation Plans and Strategies

Landscape conservation plans are typically derived from environmental data and prioritization models that attempt to protect biodiversity and natural resources cost-effectively. However, these data and maps generally do not include information such as parcel size, ownership, land use and zoning, and demographic data. Nor do they incorporate more intangible and dynamic aspects such as willing seller interest or land values and valuation trends. Therefore, as the conservation planning process evolves, land trusts supply these sorts of information to refine conservation strategies.

Landscape conservation plans seek to optimize the conservation of multiple natural resources across state to regional scales. This approach captures many but not all conservation features. Land trusts identify these gaps in regional conservation strategies, particularly as related to small-scale conservation. For example, the Florida Wildlife Corridor prioritizes large landscapes and lands connecting existing conservation areas. Accordingly, small, disconnected, and fragmented lands are not priority areas. These may nonetheless include habitats that are vulnerable to residential and commercial development or that hold imperiled species (see chapter 1). Land trusts can help identify these sorts of small-scale opportunities, fill in data gaps in landscape-level plans, and bring resources to protect these features.

Land Trusts as Facilitators: Matching Conservation Buyers and Sellers

Land trusts serve an important role in facilitating the implementation of public conservation programs, matching conservation buyers (state, federal, or local government agencies) with sellers (landowners). Land trusts shepherd projects through all stages of development: identifying matches, educating and cultivating willing sellers, evaluating eligibility, developing project applications, negotiating, and assisting with due diligence and closings. When landowners first reach out to a land trust, they may be motivated by the dual desires to protect their land while generating income. And they are looking for guidance. The land trust understands that the appropriateness of any particular program depends on the landowner's desires and objectives: Does the landowner want to sell their land outright? That generally narrows options to state or local acquisition programs.

Does the landowner want to protect agricultural lands? Perhaps a Farm Bill easement program is the best option. In later stages, land trusts help sellers navigate the transaction process, including the negotiations of conservation easement conditions and land sale terms.

As facilitators, land trusts advise and assist all parties involved in conservation transactions. The government agency (the buyer) benefits from expanded access to willing sellers through the land trust's local networks and knowledge. Landowners benefit from dealing with experienced land trust staff with firsthand knowledge of the government programs. Rather than dealing with the government, which some landowners would rather avoid, the landowner works closely with a trusted professional. The benefits of these trust relationships, often forged over years, cannot be overstated.

Land Trusts and Conservation Partnerships

Recent years have seen the emergence of public-private partnerships centered on landscape conservation goals in specific regions. Increasingly, land trusts lead these regional partnerships, consisting of government agencies, nonprofit organizations, advocacy, and other interest groups. Both formal and informal partnerships can be highly effective in leveraging resources, implementing conservation strategies, encouraging collaboration and information exchange, and accelerating the delivery of conservation programs. A good example is the "O2O" Partnership, led by North Florida Land Trust and centered on the Ocala to Osceola Wildlife Corridor (O2O; see Case Studies section).

Another example of a successful formalized conservation partnership is the Partnership for Gulf Coast Land Conservation (PGCLC), which includes 27 land trusts working in five Gulf Coast states, including Florida. The PGCLC focuses on landscape conservation within the Gulf of Mexico region and assists member land trusts with securing funding for land protection and management. This approach is particularly beneficial to smaller land trusts and conservation organizations with limited financial resources. PGCLC established a very successful Project Assistance Fund (PAF), which has supported the acquisition of 22,660 ha (56,000 ac) of land for conservation.

Public-private conservation partnerships often coalesce around public programs. Recent examples include new partnerships formed through the NRCS's Regional Conservation Partnership Program (RCPP), mentioned earlier. The RCPP provides funding for partnership development as well as direct funding for land management and protection. This partnership program has been particularly effective at leveraging federal, state, and county funding for conservation while benefiting from in-kind contributions from smaller, nongovernment organizations. Since 2017, RCPPs led by land trusts in Florida have grown from one to seven. Collectively, they have brought in nearly $70 million of federal funding for conservation programs to match a similar nonfederal contribution.

Land Trusts and Land Management—the Other Part of Conserving Florida Lands

Successful landscape conservation involves more than acquisition of land and conservation easements. Management and restoration of conservation lands is essential to protect and propagate conservation values. As a biodiversity hotspot (see chapter 1), Florida is home to diverse natural communities and fast-growing vegetation, including invasive nonnative species. Most conservation areas are too small to be self-managing, which was historically accomplished with lightning-ignited fire (Noss 2018). Today, most conservation areas must be actively managed. If neglected, conservation lands, especially those with fire-dependent ecosystems, quickly lose their conservation values.

Restoration of pine uplands is a conservation priority in Florida. Pinelands are fire-dependent ecosystems. Public conservation lands generally have dedicated land management staff and resources for fire management. However, management of private lands remains critical to reach regional conservation goals. Managing private lands, which are smaller and more likely to be proximate to development, can be problematic. As a result, land management is neither simple nor cheap, and landowners often lack financial and technical resources needed to keep up with it.

Florida public conservation lands have land management plans that are typically created with input from management staff, resource professionals, and the public. Individual landowners direct the management of their private lands. However, increasingly, regional plans and initiatives provide resources and incentives for local efforts in the context of regional goals. The America's Longleaf Restoration Initiative (ALRI undated) is a collaborative effort that supports restoration and conservation of longleaf pine ecosystems, with an ambitious goal of improving and restoring 3.2 million ha (8 million ac) of longleaf pine in the southeastern US. As part of this initiative, the National Fish and Wildlife Foundation (NFWF) and the NRCS work with public-private partnerships to deliver incentive programs and resources for private land management in support of the ALRI goals.

Florida land trusts have emerged as leaders in motivating management on private conservation lands. Several Florida land trusts lead local partnerships that are effective in reaching landowners and delivering support for conservation practices. Land trusts participate in, and often lead, regional collaboratives dedicated to accelerating land management on private lands. These "Local Implementation Teams" (LITs) are scattered throughout the southeastern US and go by various names, but they have common goals of engaging private landowners to implement land management. One is the Gulf to Atlantic Restoration Partnership (GARP), which covers a 32-county region of North Florida and South Georgia. GARP, led by Alachua Conservation Trust (ACT), brings funding and technical resources to help private landowners with land management and restoration. The GARP and similar collaboratives exemplify how regional partnerships motivate continuing efforts to manage conservation lands in private ownership and, in some cases, public lands as well.

Land trusts organize and lead landowner cooperatives dedicated to helping members manage their lands for conservation outcomes. Prescribed burning is arguably the most critical conservation land management practice in Florida. Specialized training, equipment, and workers are needed to implement prescribed burns. Two Florida land trusts have dedicated staff and resources to assist private landowners: ACT in north-central Florida and Tall Timbers Research Station and Land Conservancy north of Tallahassee. As an outgrowth of this, landowner cooperatives—prescribed burn associations—emerged to help landowners obtain the necessary training and resources. They also provide public education about the conservation benefits of prescribed fire.

In addition to technical resources, land trusts increasingly work with public programs to bring financial incentives to landowners for conservation management. The aforementioned RCPPs, led by land trusts, are an effective way to expand private landowners' access to cost-share programs. Under an RCPP agreement, land trusts work directly with private landowners to implement management plans and negotiate contracts to participate in federal cost-share programs. These are available to private landowners with or without conservation easements. However, as a by-product of incentive programs for land management, many conservation-minded landowners decide to permanently protect their lands.

Future Opportunities to Implement Landscape Conservation Plans

In 2021 the Florida legislature passed the Florida Wildlife Corridor Act (see chapter 7). This new law defines high-priority parts of the FEGN as the Florida Wildlife Corridor—a 7.3 million ha (18 million ac) area spanning the state. The Corridor represents Florida's preferred blueprint for landscape conservation. Currently, the Corridor is incomplete, with only four million ha (10 million ac) in any form of conservation land use, and many existing conservation lands in the Corridor are disconnected and isolated. The law directs the FDEP to acquire conservation lands and easements in the 3.2 million ha (8 million ac) of "opportunity areas" yet to be conserved. The law encourages voluntary conservation easements as well as novel funding sources and innovative conservation approaches such as payments for ecosystem services and other market-based approaches (Daskin and Guyton 2024). The law also calls for sustainable conservation of Florida's working ranches, farms, and forests.

As of this book's writing, public conservation funding has reached unprecedented levels in Florida, with near-record-high appropriations for Florida Forever and RFLPP (Deen 2023). The federal Inflation Reduction Act of 2022 allocated an unprecedented $20 billion to federal conservation programs, which fuels tremendous growth of federal conservation and partnership programs. County and city governments continue to propose and enact programs dedicated to land protection, particularly in northeast, Central, and southwest Florida.

Demand for land conservation programs continues to grow in Florida. Evidence of this demand includes the long waiting list of landowners seeking to sell agricultural easements through the RFLPP and increasing demand for Florida Forever and federal easement programs. Demand is also apparent as voters and local government officials propose and approve initiatives to fund land conservation. In short, public funding supply is high, and so is demand for these programs. Coupled with the state mandate to conserve the Florida Wildlife Corridor, these initiatives present an opportunity for significant advances in landscape-scale conservation over the coming years.

The question is often asked: Is it possible to protect all (or even most) of the land needed to complete the Florida Wildlife Corridor? Or even less ambitious landscape conservation plans? It is difficult to imagine the amount of funding and effort needed for this, even in light of recent developments. To achieve

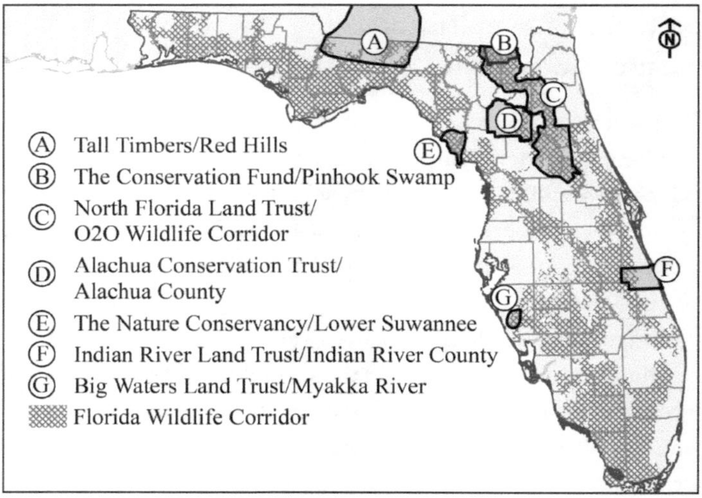

FIGURE 9.2. Map of case study areas in Florida, in the context of the Florida Wildlife Corridor. Letters denote Florida land trusts corresponding to focal areas as described in the text. Note: the shaded areas do not represent land trust operating areas, nor are all Florida land trusts shown. These are select focal areas and land trusts mentioned in the case studies.

these lofty goals, conservation practitioners will need to bring all their tools and expertise to the table. This effort will require novel and creative approaches to land conservation in addition to traditional methods. These include new approaches to leveraging funding and incentives for private lands conservation and management.

As conservationists, we are fortunate to live and work in a state replete with the science, data, and knowledge needed for effective and targeted landscape conservation—and the political will to accomplish it. Florida's landscape conservation plans help guide effective conservation, and recent successes of the Florida Wildlife Corridor have shown the benefits of well-founded planning, branding, and advocacy. However, the best-laid plans are only as good as their implementation, and the job ahead is daunting. Fortunately, Florida land trusts are well positioned to spearhead the partnerships, advocacy, outreach, and the new and creative conservation approaches needed to assist with this tremendous effort.

Case Studies of Florida Land Trust Work in Specific Landscapes

We conclude this chapter with a few case studies. While it isn't possible to describe the good work of each and every Florida land trust, these case studies (fig. 9.2) illustrate the ways in which land trusts contribute to landscape conservation.

THE NATURE CONSERVANCY

The Nature Conservancy (TNC) was an early leader of Florida landscape conservation efforts. In 1977, TNC facilitated one of the state's first large private conservation land acquisitions: a 4,800 ha (12,000 ac) land donation in the Big Bend region of Florida's Gulf Coast. This was soon followed by the state acquisition of another 32,000 ha (80,000 ac). These lands eventually became the Lower Suwannee National Wildlife Refuge, managed by the US Fish and Wildlife Service. Today, this region remains one of the nation's least developed coastlines, protected by large and connected public lands. These early land protection projects led by TNC and other organizations helped shape Florida landscape conservation for decades to follow.

THE CONSERVATION FUND

The Conservation Fund (TCF) is a national land trust, responsible for some of the largest land protection projects in Florida. In partnership with other organizations, TCF was a driving force for the establishment of the Florida Greenways Commission in the 1990s, one of the early advocates for landscape conservation in Florida. Through its land protection efforts, TCF has protected tens of thousands of acres within the Florida Wildlife Corridor. Between 2013 and 2023, TCF acquired over 9,700 ha (24,000 ac) and conveyed them to the US Forest Service via the federal Land and Water Conservation Fund. Today, these lands are part of the Osceola National Forest and protect much of the Pinhook Swamp and Greater Okefenokee landscape, one of the priority areas of the Florida Wildlife Corridor since the 1980s (Noss and Harris 1986). As a nimble national land trust with dedicated resources, TCF was able to act quickly to purchase these lands on the open market, demonstrating how land trusts with resources can acquire and hold conservation properties until public funding becomes available.

TALL TIMBERS

Tall Timbers is a regional land trust near Tallahassee dedicated to conservation of the Red Hills region along the Florida-Georgia border. This region covers nearly 162,000 ha (400,000 ac) characterized by a mature longleaf pine ecosystem and has a long history of private landowner conservation, rooted in its tradition of quail hunting from the early twentieth century. Beginning in the early 1990s, Tall Timbers worked with a coalition of conservation-minded landowners and philanthropists to complete some of the first and largest donated conservation easements in Florida and Georgia. The magnitude of this effort was remarkable, particularly at a time when conservation easements were an innovative but largely unproven conservation tool. Fortunately, Red Hills landowners were willing to make significant donations to perpetually restrict their lands for conservation. In subsequent years, other landowners followed their example. Today, the Red Hills region contains one of the best remaining longleaf pine landscapes in the Southeast, and nearly half of the region is permanently protected, much of it with donated conservation easements held by Tall Timbers. This is an example of successful landscape-scale conservation borne of early and dedicated philanthropic efforts, a trusted local land trust, and the region's strong land ethic.

ALACHUA CONSERVATION TRUST

Alachua Conservation Trust (ACT) is one of Florida's oldest land trusts, working in 22 North Florida counties. ACT is largely responsible for one of Florida's most durable and successful county conservation programs: Alachua County Forever (ACF). This program has protected over 13,769 ha (34,000 ac), including large and connected natural landscapes east of Gainesville and along the Santa Fe River Corridor to the north and west. ACT spearheaded the 2000 bond referendum that generated $29 million for conservation through ACF. In 2008, 2016, and again in 2022, ACT led the effort to renew ACF through the passage of three more ballot initiatives. In total, ACT facilitated more conservation ballot initiatives than any other local Florida land trust, transforming the landscape and future of a rapidly growing Florida county. Consistent voter approval of ballot initiatives demonstrates community and political support for regional conservation. ACT has been a pioneer in novel approaches to conservation and conservation partnerships. In 2006, ACT and the University of Florida Conservation Clinic (an experiential law school program) developed a mechanism to permanently protect parks from being sold or converted to other uses, leading to voter approval of the Registry of Protected Public Places program.

NORTH FLORIDA LAND TRUST AND O2O PARTNERSHIP

North Florida Land Trust (NFLT) and the O2O Partnership provide an excellent example of how a public-private collaborative works to implement landscape conservation. The Ocala to Osceola Wildlife Corridor (O2O) is a 648,000 ha (1.6 million ac) section of the Florida Wildlife Corridor in northeast Florida connecting North Florida's largest conservation landscapes. Led by NFLT, the O2O Partnership includes public agencies and private organizations dedicated to building the wildlife corridor to safeguard environmentally sensitive lands, working and recreational lands, and military installations, with a goal of 57,000 ha (140,000 ac) protected by mid-century.

In 2018, NFLT partnered with the US Army and Navy through the Readiness and Environmental Protection Integration (REPI) program. With REPI support, NFLT was able to acquire lands buffering installations to avoid conflicts with the military's training mission. This led to more public and private support for the O2O. The state designation of the Florida Wildlife Corridor in 2021 brought a record state budget appropriation of $400 million for land protection in the O2O. Fortunately, due to their rich experience, NFLT and the O2O Partnership were prepared for this windfall and are working with state agencies to accelerate conservation in the O2O corridor. The O2O demonstrates the power of partnerships and is a model of Florida's collaborative landscape conservation efforts.

INDIAN RIVER LAND TRUST

The Indian River Land Trust (IRLT) is a leader in partnering with local governments to provide recreational access through public trails on conservation lands prioritized by the state. IRLT has raised millions of dollars in private donations to purchase dozens of properties along the Indian River Lagoon (a state aquatic preserve and Outstanding Florida Water). Under the auspices of the Indian River Lagoon Blueway, IRLT has leveraged public dollars to construct two much-needed multiuse trail systems on some of this conserved land. Land trust staff coordinated the design and engineering of the trail systems, which include boardwalks and observation platforms. The trails seamlessly traverse parcels owned by both the county and the land trust.

BIG WATERS LAND TRUST

The Big Waters Land Trust (BWLT, formerly Conservation Foundation for the Gulf Coast) is based in southwestern Florida. The Myakka River Corridor is one of its focal regions, including 53,000 ha (130,000 ac) of protected lands surrounding the Myakka River, a state-designated Wild and Scenic River—the result of nearly 100 years of conservation efforts. In recent years, BWLT helped Myakka River landowners meet their land conservation goals by facilitating access to public and private funding programs, resulting in an additional 7,000 ha (17,000 ac) of land protected in the parts of the watershed under the greatest threat from development.

Working with programs at all levels of government, BWLT addressed specific needs of each property and owner, including a 216 ha (534 ac) NRCS Wetlands Reserve Easement (WRE) on a hunting property, a 463 ha (1,143 ac) working ranch protected with a water management district conservation easement, and a 77 ha (191 ac) nonprofit education center protected with a conservation easement held by Sarasota County. The organization also helped Sarasota County acquire a 97 ha (240 ac) public preserve and several donated conservation easements. The work of BWLT in the Myakka River watershed exemplifies its creative approach in matching conservation programs to fit the needs of different landowners and land types.

Literature Cited

America's Longleaf Restoration Initiative (ALRI). Undated. https://americaslongleaf.org. Accessed September 16, 2024.

Archbold Biological Station. Undated. https://www.archbold-station.org/projects/easements. Accessed September 9, 2024.

Bray, Z. A. 2010. Reconciling development and natural beauty: The promise and dilemma of conservation easements. Harvard Environmental Law Review 2010:119–177.

Byers, E., and K. M. Ponte. 2005. The Conservation Easement Handbook, 2nd edition. Land Trust Alliance, Washington, DC.

Cheever, F., and N. A. McLaughlin. 2015. An introduction to conservation easements in the United States: A simple concept and a complicated mosaic of law. University of Utah College of Law Research Paper 130. Salt Lake City.

Daskin, J., and J. Guyton. 2024. To conserve large landscapes like the Florida Wildlife Corridor, payments for ecosystem services could be key (commentary). June

19, 2024. https://news.mongabay.com/2024/06/to-conserve-large-landscapes-like-the-florida-wildlife-corridor-payments-for-ecosystem-services-could-be-key-commentary. Accessed September 22, 2024.

Deen, T. 2023. Conservation land got a huge boost in Florida's new budget. The Invading Sea, July 11, 2023. https://www.theinvadingsea.com/2023/07/11/conservation-land-florida-budget-wildlife-corridor-desantis-legislature/. Accessed September 11, 2024.

Dyckman, C. S., C. McMahan, A. T. Overby, N. Fouch, S. Ogletree, S. W. Self, D. L. White, M. Lauria, and R. F. Baldwin. 2024. Unpacking conservation easements' assessed land use designations and their implications for realizing biodiversity protection. Conservation Science and Practice 6:e13130.

Fishburn, I. S., P. Kareiva, K. J. Gaston, and P. R. Armsworth. 2009. The growth of easements as a conservation tool. PLOS One 4(3): e4996.

Florida Department of Agriculture and Consumer Services. Undated. https://www.fdacs.gov/Consumer-Resources/Protect-Our-Environment/Rural-and-Family-Lands-Protection-Program. Accessed September 9, 2024.

Florida Fish and Wildlife Conservation Commission (FWC). Undated. Florida's State Wildlife Action Plan (SWAP). https://myfwc.com/conservation/special-initiatives/swap/. Accessed September 16, 2024.

Florida Natural Areas Inventory (FNAI). Undated. Conservation lands. https://www.fnai.org/conslands/conservation-lands Accessed September 9, 2024.

Halperin, D. I. 2011. Incentives for conservation easements: The charitable deduction or a better way. Law and Contemporary Problems 74:29–50.

Henderson, C. 2022. Forces of Nature: A History of Florida Land Conservation. University Press of Florida, Gainesville.

Hudson, M. 2020. Nassau County Conservation Lands Acquisition and Management Manual and Conservation Plan. Nassau County Department of Planning and Economic Opportunity. https://www.nassaucountyfl.com/1283/Conservation-Land-Acquisition-Management. Accessed September 11, 2024.

Land Trust Alliance. Undated. https://landtrustalliance.org/ Accessed September 9, 2024.

Lindstrom, T. 2019. Six important developments for conservation easements. Virginia Lawyer 68(3):30–33.

Looney, A. 2017. Charitable contributions of conservation easements. Tax Policy Center (a joint venture of the Urban Institute and the Brookings Institution).

McLaughlin, N. A. 2020. Trying times: Conservation easements and federal tax law. University of Utah, S. J. Quinney College of Law, Salt Lake City.

McLaughlin, N. A. 2024. Keeping the perpetual in Florida's conservation easements. FIU Law Review 18:347–383. https://dx.doi.org/10.25148/lawrev.18.2.9. Accessed September 11, 2024.

Noss, R. F. 2018. Fire Ecology of Florida and the Southeastern Coastal Plain. University Press of Florida, Gainesville.

Noss, R. F., C. Carroll, K. Vance-Borland, and G. Wuerthner. 2002. A multicriteria assessment of the irreplaceability and vulnerability of sites in the Greater Yellowstone Ecosystem. Conservation Biology 16:895–908.

Noss, R. F., and L. D. Harris. 1986. Nodes, networks, and MUM's: Preserving diversity at all scales. Environmental Management 10:299–309.

10

Florida's Paradoxical Influence on Wildlands Protection and Restoration

John Davis, Jodi Hilty, Harvey Locke, Ron Sutherland, Charles Chester, Sonam Wangdi, and Katie Shepard

Introduction

Life is full of paradoxes, and some of them are sweet. Among the sweetest of recent times is the wildly disproportionate influence Florida has had in regional to continental-scale conservation. In this chapter we explore how it has influenced large-scale conservation in North America and beyond.

Conservation approaches need to be fit for purpose. They need to start by being grounded in present realities even if their goal is improvement. The world is not in uniform condition: some areas are still very wild, some are horribly degraded, and some are in between. Each requires a suite of conservation approaches appropriate to these starting points. At a global scale the terrestrial world can be assessed on those three conditions (Locke et al. 2019).

About 26% of the world's land is large wild areas. Here the focus is on maintaining wild conditions, preventing further loss, and ensuring large-scale natural processes like free-flowing rivers and forest-generated rainfall continue. Such areas are inherently interconnected.

At the other end of the scale, about 17% of land is cities and farms, which are half- to fully transformed by humans. It is in these areas that most people live and most food calories are produced. Natural processes are severely disrupted, and large carnivores and large-bodied wild grazers have been eliminated. Yet they can harbor populations of rare and endemic species and high richness of small-bodied animals and plants. Here, the focus of conservation is to retain all fragments, keep pollutants out of water courses and groundwater, secure remaining populations of native species, and restore natural processes and connectivity to the extent possible.

Most of the world's land, about 55%, is in shared landscapes which vary from half transformed to few significant impacts (see Locke et al. 2019 for methods). Land uses in shared landscapes vary from national parks and wilderness to logging, mining, hydro dams, and small farms with grazing. These shared landscapes are bisected by roads and rail lines. Shared landscapes often retain most of their native species (sometimes including large carnivores and large-bodied grazers) but in depleted numbers, and their habitats are often fragmented, sometimes to the point of isolation. It is in this shared landscape condition that the Florida model was developed and has achieved global influence.

Here we offer perspectives on the various ways the Florida model has influenced conservation. John Davis explores how it has played a key role in shaping the rewilding movement in North America and how its science-based approach intersects with wilderness advocacy. Harvey Locke and Sonam Wangdi discuss its influence at a global scale and connection to the Nature Needs Half movement with a special focus on Bhutan, which has protected over half the country in an interconnected way. Jodi Hilty and Charles Chester describe the Florida model's connection to the Yellowstone to Yukon Conservation Initiative's approach—where all three landscape conditions are present—with a focus on how it intersects with Indigenous-led conservation. Last, Ron Sutherland describes how the Conservation Area Design (CAD) approach pursued by the Wildlands Network (once known as the Wildlands Project) applies the Florida model in the mid-latitude regions of North America.

The Influence of the Florida Model on Regional Wildland Network Designs

John Davis: The core/corridor/buffer concept that Larry Harris and Reed Noss (e.g., Noss and Harris 1986) first articulated became the template that Wildlands Network leaders, particularly Dave Foreman and Michael Soulé, and visionaries with other conservation groups, too, used in their conservation designs. Indeed, atop the list of books many North American rewilding proponents read are *Saving Nature's Legacy* by Reed Noss and Allen Cooperrider (1994) and *Rewilding North America* by Noss's close colleague Dave Foreman (2004). Foreman's landmark book is perhaps the closest thing yet to a blueprint for how the midpart of North America can implement the goal of protecting at least half of the world—known variously as Nature Needs Half (Locke 2013) or Half-Earth (Wilson 2016).

In *Rewilding North America*, Foreman urged conservationists to give special attention to four megalinkages: the Atlantic/Appalachian, or Eastern; the Spine of the Continent, or Rocky Mountain; the Pacific Crest and Coast; and the Boreal, across much of Canada and Alaska. Followers of Dave Foreman later convinced him to replace the term "megalinkages" (which, as several readers wryly noted, sounded like giant sausages) with the term "continental wildways." We also convinced Dave to do a second edition of *Rewilding North America*, and to propose in it Gulf Coast and Great Plains Wildways, too. The Gulf Coast Wildway would encompass much of the Coastal Plain, which Noss and colleagues succeeded in getting recognized as a global biodiversity hotspot (Noss et al. 2015). More specifically, it would capture many of the *Forgotten Grasslands of the South* (Noss 2012), which Noss lucidly described in his book of that name. Sadly, Dave Foreman died before he could complete the second edition.

The regional wildlands network designs that Foreman and other Wildlands Network leaders drafted showed at a finer scale just where optimum locations for big wild cores and connecting corridors would be and how they might be buffered. In Sidebar 10.1, we list many of the wildlands network designs employing some version of the Florida model. Dave Foreman and his colleague Dave Parsons (a retired US Fish & Wildlife Service biologist who had overseen the Mexican Wolf recovery program) gave much time to the New Mexico Highlands, Sky Islands, and Southern Rockies wildlands network designs, all of which share roots with the Florida model.

Another paradox here, though, may be that as Reed Noss, Tom Hoctor, Carlos Carroll, and other conservation biologists were taking the science of regional landscape conservation planning to higher and higher standards, Dave Foreman was warning that while conservation groups should be informed by the science, they should not be limited by it. If grassroots conservationists could make a good case for the Wilderness designation of some place they loved, then that place should be protected, whether or not it showed up on scientists' maps. Foreman was ambivalent about computer models and GIS, at heart preferring good old-fashioned hand-drawn, ground-truthed wildlands proposals sketched on mylar then implemented through legal protections (like Wilderness or National Park status on public land or Forever Wild easements on private land).

Still, unlike most conservation plans—which if bold are mere visions and if timid are scarcely worth the trees consumed in the drafting—the Florida wildlands plan posited and refined by Harris, Noss, and Hoctor actually has lived up to the name "plan": it has to a large degree guided state land acquisitions (including easements) for the last few decades. Many wildlands network designs are really just visions or proposals, too seldom considered in land transactions or zoning, much less guiding state conservation policy and conservation land acquisition.

Along with the Florida model, another surprisingly successful conservation plan is the Adirondack Council's *2020 Vision* series (published in the late 1980s as booklets but now available online: Adirondack Council 2020), which gave New York State an ambitious priority list for realizing the potential of Adirondack Park, resulting in hundreds of thousands of acres more protected as Forest Preserve or by conservation easement. The conservation community would benefit from studying why the Florida conservation model and the Y2Y (discussed in the section "Yellowstone to Yukon Region") and Adirondack visions each proved uncommonly effective in wildlands advocacy. We cannot do that study here; but for now, I suggest that the Florida and New York models have guided implementation largely because they were grounded in both conservation science and field time (I had friends, now dead, who told me of the countless hours they spent ground-truthing 2020 proposals for Adirondack Park). Undoubtedly another reason for the lasting success of the Florida Forever and Adirondack 2020 proposals is that voters in Florida and New York periodically approve major state expenditures for land conservation—under both Democratic and Republican governorships.

The Florida model influenced much more than just subsequent wildlands network designs; it informed numerous books and articles. Look particularly at the titles of Island Press, which publishes many of the most influential conservation and environmental books, and you will find several books by Reed Noss himself but also many books by other scientists and conservation leaders advancing reserve designs and conservation. As well, Noss's article in the 1992 Special Issue of *Wild Earth*, "The Wildlands Project land conservation strategy" (Noss 1992), reviewed the evidence that about half of any given area studied in the US and similar regions needs to be protected to conserve its biodiversity, and it was also a recipe for effective conservation designs. The first scientific explanation in the popular literature for rewilding was the *Wild Earth* article "Rewilding and Biodiversity: Complementary Goals for Continental Conservation" by Michael Soulé and Reed Noss (Soulé and Noss 1998).

In sum, the Florida conservation model has inspired and informed bountiful wildlands and rewilding plans and publications, especially in temperate North America but globally as well. To the extent that Nature Needs Half goals are achieved, it will be largely due to visionaries in Florida.

The Influence of the Florida Conservation Model of Large Landscape Conservation

Harvey Locke and Sonam Wangdi: The Florida Conservation model of cores and corridors centered around the life needs of an endangered large carnivore came at a pivotal moment in conservation history. By the mid-1980s, island biogeography theory had moved from the observation of actual islands (MacArthur and Wilson 1967) to terrestrial "island" national parks encircled by incompatible development (Newmark 1985). Earlier, concerns about incompatible development on the boundary of core protected areas had led to greater ecosystem thinking in the US (Leopold 1949) and globally to the creation in 1971 of UNESCO biosphere reserves that were intended to buffer protected areas (UNESCO 1974). But all of these still left out the idea of interconnections among the reserves to effectively increase their size in order to maintain viable populations of wide-ranging mammals.

The fragmented and diminishing population of the iconic Florida panther (*Puma concolor*) put a face on the problem of isolated small reserves at a statewide level. A large landscape vision that combined theory and the charismatic panther made the idea understandable, practical, and inspiring; it was the Florida model that literally put the concept of cores, corridors, and buffer zones on the map so people could more easily visualize it (Noss 1985; see chapter 2, fig. 2.1).

By 1990 this concept and associated theory were scaled up to a larger subtropical and tropical conservation vision known as *Paseo Pantera* to connect Florida across the US Gulf states and through Mexico and Central America (Vega 1994). It also led to the conceptualization in Costa Rica of a national plan for protected areas interconnected by biological corridors (Barborak et al. 1993). It was also a key part of the founding of the Wildlands Project, a vision for interconnecting Central and North America (Cenozoic Society 1992).

Large landscape conservation was an idea whose time had come, and the

NORTH AMERICAN WILDLANDS PROPOSALS AND PAPERS INFLUENCED BY THE FLORIDA MODEL

On the Rewilding Institute (TRI) website (rewilding.org):

"Selecting and Designing Protected Areas: The Early Days" by Dave Foreman: https://rewilding.org/selecting-and-designing-protected-areas-the-early-days/

"We Need Big Holistic Wilderness" by Howie Wolke: https://rewilding.org/we-need-big-holistic-wilderness/

"The Mogollon Wildlife Corridor: A Big Step Toward Half Earth and Beyond" by Kim Crumbo: https://rewilding.org/the-mogollon-wildlife-corridor-a-big-step-toward-half-earth-and-beyond/

"Recommended Wildlife Corridor Goals and Objectives," Drafted for The Rewilding Institute by Kim Crumbo, TRI Wildlands Coordinator; With input from Bob Howard, TRI; and Ron Sutherland, Wildlands Network: https://rewilding.org/recommended-wildlife-corridor-goals-and-objectives/

"America's Red Rock Wilderness Act: Moving America Closer to 30x30 and Enhancing Wildlife Connectivity" by Kim Crumbo: https://rewilding.org/americas-red-rock-wilderness-act/

(The full paper: https://suwa.org/wp-content/uploads/30x30-Report-FINAL-1.20.2021-WEB.pdf)

"Thirty By Thirty and Half Earth: Promises and Pitfalls" by Howie Wolke: https://rewilding.org/thirty-by-thirty-and-half-earth-promises-and-pitfalls/

"Connecting at All Scales: Following Scientists and Activists" by John Davis: https://rewilding.org/connecting-at-all-scales-following-scientists-and-activists/

"A Brief History of the National Wilderness Preservation System" by Dave Foreman: https://rewilding.org/a-brief-history-of-the-national-wilderness-preservation-system-2/

"New Report Offers Road(less) Map to Continental Wildlands Network" by Jamie Sayen: https://rewilding.org/new-report-offers-roadless-map-to-continental-wildlands-network/

(The full report: "Wildlands in New England: Past, Present, and Future" https://wildlandsandwoodlands.org/resources/wildlands-in-new-england/)

"Yellowstone to Uintas" by Kirk Robinson: https://rewilding.org/yellowstone-to-uintas/

"Split Rock Wildway" by John Davis (three-part series): https://rewilding.org/split-rock-wildway/

"An Ecological Approach to the Selection, Design, Prioritization, Management, and Restoration of Wilderness Areas" by Dave Foreman: https://rewilding.org/our-programs/bedrock/

"The Science Behind Continental-Scale Conservation" by The Rewilding Institute: https://rewilding.org/the-science-behind-continental-scale-conservation/

On the Wildlands Network website (wildlandsnetwork.org):

Southern Rockies Wildlands Network Vision: A Science-Based Approach to Rewilding the Southern Rockies: https://wildlandsnetwork.org/resources/southern-rockies-wildlands-network-vision

Maine Wildlands Network Vision: A Scientific Approach to Conservation Planning in Maine: https://wildlandsnetwork.org/resources/maine-wildlands-network-vision

New Mexico Highlands Wildlands Network Vision: Connecting the Sky Islands to the Southern Rockies: https://wildlandsnetwork.org/resources/new-mexico-highlands-wildlands-network-vision

From the Adirondacks to Acadia: A Wildlands Network Design for the Greater Northern Appalachians: https://wildlandsnetwork.org/resources/adirondacks-to-acadia-wildlands-network-design

Eastern Wildway Map: https://wildlandsnetwork.org/resources/eastern-wildway-map

Envisioning a Connected California: https://wildlandsnetwork.org/resources/envisioning-a-connected-california

Identifying and Prioritizing Key Habitat Connectivity Areas for the South Atlantic Region: https://wildlandsnetwork.org/resources/key-habitat-connectivity-in-the-atlantic-region

Heart of the West Conservation Plan: https://wildlandsnetwork.org/resources/heart-of-the-west-conservation-plan

Borderlands Connectivity Report: https://wildlandsnetwork.org/resources/borderlands-connectivity-report

Southeast Climate Science Center Final Connectivity Report: https://wildlandsnetwork.org/resources/southeast-csc-final-connectivity-report

In print—landscape conservation proposals and texts influenced by the Florida model:

Missing Linkages: Restoring Connectivity to the California Landscape; 2000; proceedings of a summit convened by California Wilderness Coalition, The Nature Conservancy of California, and South Coast Wildlands Project.

Noss, R. F. 1992. A Preliminary Biodiversity Conservation Plan for the Oregon Coast Range. Coast Range Association, Newport, OR.

Journal version: Noss, R. F. 1993. A conservation plan for the Oregon Coast Range: Some preliminary suggestions. Natural Areas Journal 13:276–290.

Strittholt, J. R., R. F. Noss, P. A. Frost, K. Vance-Borland, C. Carroll, and G. Heilman. 1999. A conservation assessment and science-based plan for the Klamath-Siskiyou ecoregion. Earth Designs Consultants Inc. and Conservation Biology Institute, Corvallis, OR.

Journal version: Noss, R. F., J. R. Strittholt, K. Vance-Borland, C. Carroll, and P. Frost. 1999. A conservation plan for the Klamath-Siskiyou ecoregion. Natural Areas Journal 19:392–411.

Noss, R., J. Strittholt, G. Heilman, P. Frost, and M. Sorensen. 2000. Conservation planning in the redwoods region. Pages 201–228 in R. Noss, editor. The Redwood Forest: History, Ecology, and Conservation of the Coast Redwoods. Island Press, Washington, DC.

Noss, R. F., G. Wuerthner, K. Vance-Borland, and C. Carroll. 2001. A Biological Conservation Assessment for the Greater Yellowstone Ecosystem: Report to the Greater Yellowstone Coalition. Conservation Science Inc., Corvallis, OR.

Journal version: Noss, R. F., C. Carroll, K. Vance-Borland, and G. Wuerthner. 2002. A multicriteria assessment of the irreplaceability and vulnerability of sites in the Greater Yellowstone Ecosystem. *Conservation Biology* 16:895–908.

Trombulak, S. C., and R. F. Baldwin. 2010. Landscape-scale Conservation Planning. Springer, New York.

Headwaters (of New England) proposal, originally drafted ca. 1995 by Jamie Sayen then summarized in his book CHILDREN OF THE NORTHERN FOREST: A History of the Natural and Human Communities of Northern New England (2023).

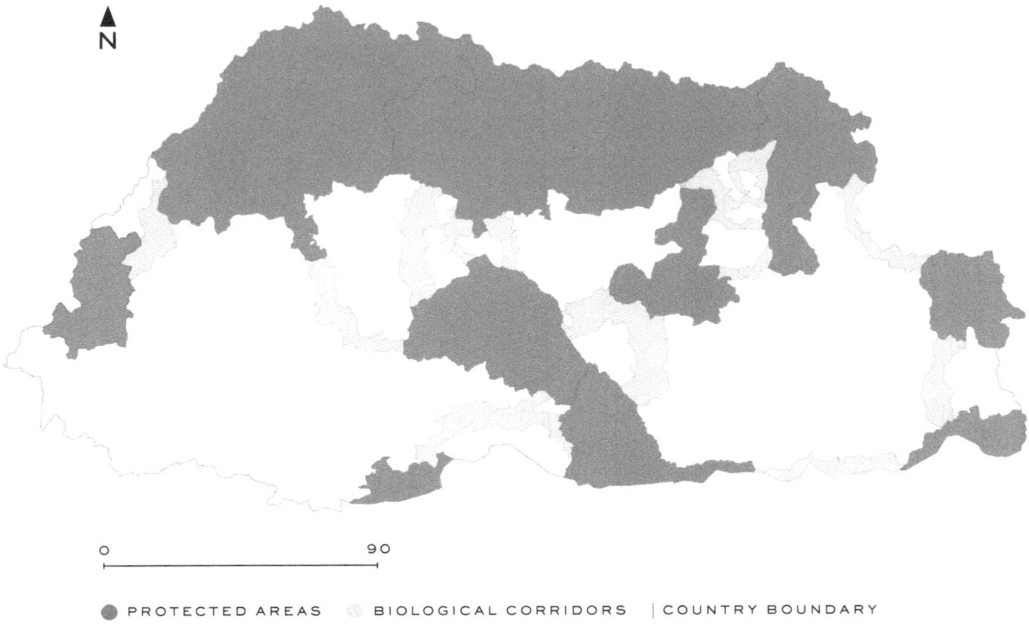

FIGURE 10.1. System of protected and conserved areas and corridors of Bhutan. The new 2024 corridor (BC 9) is shown in the upper right. Adapted from unpublished map, Department of Forests and Park Services, Ministry of Energy and Natural Resources, Bhutan National Geodatic Datum.

Florida model was an example of how to do it. It and *Paseo Pantera* were influences on the creation of the Yellowstone to Yukon Conservation Initiative. In the Northern Rockies of the US and Canada, gray wolf (*Canis lupus*) movements revealed actual travel among reserve networks (Locke 1994), validating the wide applicability of the Florida theory. Reed Noss spoke about Florida in his presentation at the founding meeting of Y2Y in December 1993 and again at its public launch in 1996. Jodi Hilty and Charles Chester elaborate on this history in the following section.

For an echo of the Florida model, we can look at the extraordinary system of interconnected protected areas created by the Kingdom of Bhutan (Bhutan Biodiversity undated). Large core areas in different ecologically representative landscapes span the vast elevational differences of this highly biodiverse country. Using the results from a nationwide tiger survey, the areas were connected by ridgeline corridors designed with tiger (*Panthera tigris*) movements in mind. In 1999 these biological corridors were bequeathed as "A Gift to Earth from the

People of Bhutan." Eight such corridors were initially designated, which increased Bhutan's protected and conserved areas to 51.4% of its country's geographical area.

Connectivity between protected areas is still a major motivation in Bhutan, with a new corridor connecting reserves in the eastern part of the country designated by Parliament in 2024. This was a result of a local biodiversity survey where a tiger was camera-trapped along with many other wild fauna and flora. With the addition of the ninth biological corridor, Bhutan now has 52% of its land managed as protected areas (fig. 10.1).

It is easy to spot the ongoing influence of the Florida model today. For example, a 2024 paper on the tigrina, a small carnivore in Brazil's highly diverse mid-Atlantic rainforest in Brazil, entitled "Prioritizing conservation areas to mitigate connectivity loss and local extinction risk of a small carnivore (*Leopardus guttulus*) in South America" (Sartor et al. 2024) might be called its grandchild, right down to the cores and corridors it identifies that resemble those on the Noss map. Although it uses a landscape genetics computer program, resistance kernels, and microsatellite data to generate its up-to-date outcomes, the basic approach of the Florida grandparent is plainly apparent even if it is not acknowledged in the references.

The thinking associated with the development of the Florida model also led to asking the question: how much needs to be in core reserves? At the time it was developed, there was talk about tripling the global protected area system to 10 or 12% (Locke 2018). But long before that, Odum and Odum suggested that about half of South Florida needs to be protected to secure what are now known as ecosystem services, and they advised planners elsewhere to take a similar approach (Odum and Odum 1972; see chapter 2). In 1993 at the founding meeting of the Yellowstone to Yukon Conservation Initiative, Reed Noss spoke of the need to conserve about half of a landscape in an interconnected way to preserve biodiversity and natural processes. Through similar thinking, in 1999 Bhutan was the first country to achieve this level of interconnected conservation. The seeds sown were a factor in the development of the global Nature Needs Half movement (Locke 2013, 2018).

The cores and corridors model has a very wide application worldwide. It is best suited to landscapes that are neither largely intact nor highly fragmented (see Introduction). This middle condition, called shared lands in the Three Global Conditions for Biodiversity Conservation and Sustainable Use, covers about 55% of the terrestrial world (Locke et al. 2019).

The influence of ideas that make sense for the world is a funny thing. They can become so self-evident that the original source is not recognized. The Florida model of appropriately buffered core areas connected by corridors, designed to support wide-ranging carnivores and all the species that come with them, was the first example of what is now one of the fundamental concepts in modern conservation practice. To paraphrase Thomas Jefferson's writing in the US Declaration of Independence, in the world of conservation, the Florida conservation model has almost become a truth we hold to be self-evident.

Yellowstone to Yukon Region

Jodi Hilty and Charles Chester: In the early 1990s a group of conservationists and scientists agreed on a vision to advance an interconnected system of wild lands and waters stretching from Yellowstone to Yukon (Y2Y), harmonizing the needs of people with those of nature, to ensure that large carnivores and hooved animals would be able to roam widely and thrive into the future (Chester 2006; Hilty and Zenkewich 2022). This early movement included both American and Canadian scientists and conservationists who realized that the science showed that while protected areas are important, they are not adequate in isolation—instead, we need a network of protected areas that are connected across what we know today is the wildest large mountain region in the world (Theobald et al. 2024). Achieving a vision of this enormity across two countries requires science, community work, and enabling policies (Hebblewhite et al. 2021). To date, over 700 different entities—ranging from businesses, governments (federal, regional, and Indigenous), communities, scientists, and nonprofits—have engaged at different times to help advance this vision (M. Strebel, personal communication). Today, the Yellowstone to Yukon Conservation Initiative, a joint effort between a US nonprofit organization and a Canadian public charity, helps ensure that key collaborative work continues to advance at the scale that nature needs.

Given the huge scope of the effort, many have wondered whether working at such a scale can be effective. Scientists recently examined whether the Yellowstone to Yukon (Y2Y) vision, in fact, was making a difference. Protected areas have increased by 80% in the first 25 years, which is double the rate of protected area growth across North America as a whole (Hebblewhite et al. 2021). At least a quarter of all the area protected in the Y2Y region is managed or comanaged by Indigenous people (Hilty and Zenkewich 2022). Likewise, few wildlife road-crossing structures existed in 1993, but today the Y2Y region has at least 130 designated wildlife underpasses and overpasses and associated fencing to help keep wildlife safely connected across busy roads (Hebblewhite et al. 2021; updated analyses, K. Zenkevich). With additional crossings agreed upon and being built, and promising higher-level policy and funding, consideration of safe wildlife passage is moving toward being normalized (e.g., Goldfarb 2023; Montana Fish, Wildlife, and Parks 2023). Other progress includes the advancement of voluntary private land conservation, especially in critical corridors for wildlife, and efforts to ensure that communities and wildlife can successfully coexist and that some endangered species, such as US grizzly bears (*Ursus arctos horribilis*), are recovering. Also, other large landscape movements around the world have taken inspiration and insight from Y2Y (Hebblewhite et al. 2021).

A theme from the beginning of Y2Y is engaging with Indigenous peoples to advance shared conservation priorities, and today the starting point for conservation is working with Indigenous people through a reconciliation framework. In Canada, this is based in public policy, and all conservation is advancing across the country with such an approach (Weston and Greenfield 2022). The Y2Y region overlaps with at least 75 Indigenous territories of First Nations, Inuit, Métis, and Native Americans. In Canada, Indigenous leadership driving forward protected area creation has accelerated in recent decades. Two prominent examples are the 120,000 km2 (29,652,646 ac) Nahanni and the 4,895 km^2 (1.19 million ac) Náátsʼįhchʼoh National Park Reserves, which became comanaged by Parks Canada and First Nations in 2009 and 2012, respectively.

Likewise, more recent protected areas in the Y2Y region are with Indigenous leadership, such as ca. 56,656 km^2 (14 million ac) of new protected areas agreed upon in signed conservation agreements between the federal, regional, and First Nations governments. These include the Peel watershed land-use plan in Yukon Territory; the Peace River Caribou Agreement to restore and rewild caribou (*Rangifer tarandus*) in northeastern British Columbia (BC); and Qat'muk, the home of the Grizzly Bear Spirit in the upper Columbia River in southeastern British Columbia (Y2YCI 2021). The collective governments are currently working toward implementation of these agreements, including on-the-ground designation.

New efforts have gained publicity and traction. For example, three Indigenous-led conservation initiatives in BC have been announced, resulting in the intended protection of 17,806 km^2 (4.4 million ac). These include a historic implementation agreement between BC and the Blueberry River First Nations; a Conservancy in the Incomappleux of BC's rare inland temperate rainforest in the Upper Columbia region; and the Taku River Tlingit declaration specifying their plan for an Indigenous and Conserved Protected Area in the Taku River watershed. In addition, the Dene K'éh Kusān is a protected areas vision by the Kaska Dena on ancestral lands and is BC's largest intact landscape, of 39,000 km^2 (9.6 million ac) (Dena Kayeh Institute, undated).

The Y2Y portion of the region in the United States also holds examples of Indigenous leadership in conservation. Both the Wind River Reservation in Wyoming as well as the Confederated Salish and Kootenai Tribes in Montana established designated tribal wilderness areas on their respective reservations in the 1930s and 1982, respectively. The Confederated Salish and Kootenai also determined the installation of one of the most progressive sets of wildlife road-crossing structures in the United States on a road bisecting their reservation. Phase I of this work included 42 crossing structures, including one overpass, within a 90 km stretch of road, and now at least 22,000 animals use these crossings annually (Christy and DiGirolamo 2022). The tribe is now initiating the second phase of the project, starting with a major structure to be installed where at least 11 grizzly bears have been killed in the last five years (Sagner 2023). Likewise, the Blackfeet in Montana led efforts with the conservation community to retire oil and gas leases in the Badger-Two Medicine region (Lundquist 2019). While not at the same level of commitment as Canada, the US federal government has signaled recently that Indigenous engagement in conservation is important.

The following are vignettes of projects that serve as examples of rewilding, a subset of which highlight increasing Indigenous leadership in conservation.

LARGE CARNIVORE REWILDING IN THE SOUTHERN REGION OF Y2Y

During the inception of the Y2Y vision in the mid-1990s, many of the individuals at the founding meetings were concerned with the already extensive range loss across North America for both large carnivores as well as hooved animals (Harvey 1998). As a mapping exercise by Laliberte and Ripple (2004) demonstrated, one of the last places where most of these animals still roam is within the Y2Y region.

The northern part of Y2Y is one of the wildest regions left on Earth, but the farther south one looks, the more fragmented it becomes. There have been a few extirpations of large carnivores and grazers. Also, some carnivore and large-grazer populations have been isolated in the Greater Yellowstone Ecosystem and the Cabinet Mountains of northwest Montana. Of particular concern is the fragmenting effect of major highways and associated development that bisect the southern part of the Y2Y region.

Worldwide, applied research has increasingly shown that the loss of large predators leads to a cascade of ecological impacts affecting multiple parts of ecosystems (Smith et al. 2020). The Greater Yellowstone Area has been a de facto natural laboratory for a plethora of research on such trophic cascades due to the extinction of several large carnivores from all or parts of the ecosystem in the early 1900s and the subsequent restoration or rewilding of carnivores back into the system. Mountain lions (*Puma concolor*) and wolverines (*Gulo gulo*) naturally reoccupied the Greater Yellowstone Ecosystem, grizzly bears (*Ursus arctos*) recovered, and gray wolves (*Canis lupus*) were reintroduced, both with a great deal of human intervention (Yellowstone Science 1994; McKelvey et al. 2014). The impacts of these recoveries are well documented and often profound (Smith et al. 2020). Less publicized but also important was the wolf reintroduction to the Idaho wildlands complex by the Nez Perce, exercising their treaty right at approximately the same time (Nez Perce Wildlife Division undated).

RESTORATION OF HABITAT AND WILDLIFE FROM THE TRANSBOUNDARY CABINET PURCELL MOUNTAIN REGION TO CAMAS TO CONDORS CORRIDOR THROUGH COLLABORATIVES IN THE PACIFIC NORTHWEST

The Cabinet Purcell Mountain Corridor Project is illustrative of how more than 50 partners working on a shared goal can make significant progress (Proctor et al. 2018). In the early 1990s, grizzly bears in the transboundary Cabinet Purcell Mountain region of Montana, Idaho, and British Columbia were showing signs of isolating into smaller populations, with one group as low as 10 individuals in Montana's Cabinet Yaak Mountains. Grizzly bear science helped to prioritize where key core habitat and connectivity zones should be protected and restored. This collaboration led to the securing of core habitat through road removal projects on US Forest Service land, securing and restoring key corridors, and massive coexistence awareness and implementation. Recent research shows increased connectivity between previously isolated populations (Proctor et al. 2018; Hilty et al. 2019). Likewise, the Cabinet Yaak population of grizzlies increased to over 60 individuals.

The Nez Perce Tribe, which engaged in the Cabinet Purcell partnership, is today leading the Camas to Condors Corridor Project (NPTWRD 2019). Modeled to a degree on the Cabinet Purcell partnership, this initiative seeks connectivity for wildlife during this time of climate change and restoration of cultural relationships with nature. Many projects are underway, and planning for a potential condor (*Gymnogyps californianus*) reintroduction is in early stages.

BISON RESTORATION

Bison (*Bison bison*) is a megaherbivore keystone species that literally shapes the ecosystem it occupies. The history of their collapse across North America is generally well known, and three key places where bison restoration is advancing today are in the Y2Y region. The first is on the northern and western edges of Yellowstone National Park, a region where bison are finally allowed to leave the park, although into a still limited buffer zone (White et al. 2011; National Park Service 2018). In the second example, in 2017 an experimental population of bison was restored to the northern reaches of Canada's Banff National Park. Now there are approximately 80 bison that roam freely in the park (Parks Canada 2022). However, like Yellowstone, bison are prevented from leaving the park, except for a small buffer area on nonpark public lands. Of note, neither park fully allows for the restoration of bison's main historical predator, humans (a.k.a. Indigenous people), an issue yet to be resolved.

As the third example, the Blackfoot Confederacy, Indigenous peoples living along the Montana-Alberta border, have been advancing the Iinnii (meaning bison) Initiative since 2009. This effort seeks to conserve traditional lands, maintain Blackfeet culture, and enable bison to return (Blackfoot Nation 2020). In June 2023, more than 40 bison were released to roam free in the Chief Mountain area east of Glacier National Park (Scott 2023). Someday, bison may once again roam across broader Blackfoot Confederacy territorial lands in Canada and the United States.

The vision for the Y2Y region is to connect, restore, and protect the region so that both people and nature can thrive, and accomplishing it means protecting extant nature as well as rewilding and restoring key ecosystems and species and maintaining and restoring human connectivity with nature. The urgency of this work has become existential with the dual global crisis of biodiversity loss and climate change threatening all life on Earth, and the imperative to demonstrate that threats can and must be tackled simultaneously and provide the world a model to move forward.

Influence of Florida Wildlands Designs on Wildlands Network Mapping 2010–2024

Ron Sutherland: Wildlands Network (WN) is a nonprofit conservation group whose mission is to reconnect, restore, and rewild nature across North America. Founded in 1991 as the North American Wilderness Recovery Strategy and later changing its name to the Wildlands Project, the organization shares deep organic roots with the Florida conservation mapping completed in the 1980s by Reed Noss and Larry Harris. At the most fundamental and obvious level, this is because Reed Noss went on to work for the Wildlands Project as its first science director (1992–1996) and chief scientist (2002–2004), and he brought with him the same vision for large-scale habitat connectivity and wilderness recovery that he had pushed for in Florida earlier. We refer readers to the checklist Reed published in 2003 in the journal *Conservation Biology* (Noss 2003), a manual for completing new wildlands network designs of the sort that would become our trademark in the 1990s and 2000s.

When I joined Wildlands Network in 2010, the organization had recently published its last (for the moment) regional Wildlands Network Design for the heavily forested Northern Appalachian ecoregion. I was given marching orders by the Wildlands board (especially my PhD adviser, John Terborgh), WN Executive Director Margo McKnight, and WN Eastern Wildway Director Conrad Reining, that my job was to complete something similar for the rich biodiversity of the Southeast.

Because of the intensive mapping and equally intensive land conservation follow-through in Florida of which I was already aware (as a graduate student), and because of my home base in central North Carolina, I decided to focus my attention on the South Atlantic states, to develop a plan for connecting habitats from North Florida to southern Virginia and west to the Appalachian Mountains. This dovetailed nicely with the geography of the newly formed South Atlantic Landscape Conservation Cooperative (SALCC), and we were able to obtain funding from the SALCC (which has now morphed into the Southeast Conservation Adaptation Strategy, or SECAS) to create habitat connectivity models for a set of seven terrestrial species. Those species included Florida panther, red wolf (*Canis rufus*), black bear (*Ursus americanus*), eastern diamondback rattlesnake (*Crotalus adamanteus*), timber rattlesnake (*Crotalus horridus*), pine snake (*Pituophis melanoleucus*), and box turtle (*Terrapene carolina*), a group that by design spanned a huge range of space use and movement requirements. Our hope was that the models would support the mapping efforts of the SALCC, ably led by Rua Mordecai. The SALCC Blueprint, which became the basis and inspiration for the SECAS Blueprint that Rua now maintains, is essentially a federalized version of the same sort of hub-and-spoke Wildlands Network Design that Reed Noss pioneered. We also obtained funding from the Southeast Climate Science Center, another federal entity whose support allowed us to extend a few of the models west to Texas and north to West Virginia.

I realized once I dove into the Southeast mapping work that I was standing on the shoulders of not only Reed Noss and Larry Harris but also their protégé, Tom Hoctor, who had extended the 1980s maps into a much more formal Florida Ecological Greenways Network, a version of which is still the official plan today (see chapter 7). Tom didn't stop with Florida, however; he and his collaborators (including lead author Peggy Carr) extended his mapping approach to the entire Southeast with the Southeast Ecological Framework, published by the US Environmental Protection Agency (Carr et al. 2002).

After we finished the connectivity models for the SALCC and Southeast Climate Science Center (with help from numerous academic collaborators, including Paul Leonard, Rob Baldwin, Nick Haddad, Jen Costanza, and others), we were approached by the Blue Ridge Forever Coalition, a group of conservation land trusts serving western North Carolina. The coalition wanted us to help them prioritize habitat connectivity options for that part of the state, so we delivered to them a fine-scale hub-and-spoke network that was built from all of the relevant model layers (e.g., black bear, timber rattlesnake) we had in our arsenal. The mapping had a new twist, though, inspired directly by the hand-drawn mapping of Noss and his counterpart at Wildlands Project, Dave Foreman. After we stacked the deeply pixelated GIS layers from the connectivity models, we drew in the corridors by hand, using human intelligence to break through the complexities of real mountain landscapes and simplify the connectivity options into discrete pathways that would lend themselves to on-the-ground conservation.

Trying to make sure the science-based maps were implemented was an urgent theme of our connectivity work that was borrowed completely from Harris's, Noss's, and Hoctor's legacy in Florida, where billions of dollars have been fruitfully spent over the course of decades to try to implement their fundamental plan. At least one land trust, the robust and well-funded Southern Appalachian Highlands Conservancy, has made excellent strides protecting several of the key corridors we identified in the North Carolina mountains, though of course there is more work to be done.

The same combination of science, art, intuition, and audacity that was pioneered by Reed Noss and used by us in the Southern Appalachians was soon brought to bear on a much more ambitious Wildlands Network project. For years, ever since intrepid eco-adventurer John Davis had coined the term, Wildlands Network had focused on presenting the vision of continental "wildways," great swooping arcs across North America that symbolized the most critical connections that needed to be maintained or restored. These largely included the mountains—the Sierra and Cascades on the Pacific, the Rockies, and the Appalachians. The "swoop map" of wildways illustrated the concept fairly well, but what the map didn't do was operationalize those plans and make them real. By the time I started at Wildlands Network, they had accumulated a stack of hub-and-spoke designs for subregions of the Western Wildway (referred to by Michael Soulé as the Spine of the Continent), but I wondered whether we could skip the region-by-region approach for the east and create one master map of the Eastern Wildway (or as I referred to it jokingly in a talk at a landscape ecology conference, "one map to rule them all!").

The goal of a continental-scale wildway was to show how an entire side of temperate North America (in this case) could be reconnected, restored, and rewilded. It would provide both an ambitious blueprint and an inspirational vision that could keep land trusts and agency conservationists busy for decades trying to stitch the continent's natural wonders back together again into a rapidly healing superorganism. We wanted the Eastern Wildway to be science-based, but we didn't want it to suffer from the same pixelated look that was common to many raster-based mapping schemes at the time. So again, borrowing from Noss, Harris, Foreman, and others, we had the audacity to use human intelligence to draw, essentially by hand, cores and corridors from Florida to the Mississippi River and north to Quebec and Ontario. The GIS base layers we used to draw from included all the best science products available in the mid-2010s, but the one we started with (and largely left untouched, as there was no real way of improving on it) was the Florida Ecological Greenways Network.

Putting together the science layers took months, but actually drawing the first draft of the Eastern Wildway map only took me a few weeks of rapid-fire scrolling and clicking, and suddenly the first draft of the Wildway was ready to be shared with staff on February 27, 2017. We strove when drawing the Eastern Wildway to hit the Nature Needs Half or Half-Earth goal of protecting at least 50%. We allowed ourselves the freedom to dream big when crafting the Eastern Wildway, looking beyond just the most pristine habitats that were left and asking ourselves, what could still be restored, rewilded, and brought back to life as part of the network? We thus included many areas that were degraded in various ways, such as pine plantations that could be restored to longleaf pine savannas and farm fields that could be restored to native forests and grasslands, and those inclusions make the Eastern Wildway unique among its peers in the conservation mapping world.

One last connection between the Florida mapping and the Eastern Wildway that I should highlight is the use of an actual artist to take the GIS-derived vision and turn it into a work of compelling beauty. We saw the beautiful map of the Florida Wildlife Corridor by Mike Reagan and knew instantly that we would want something similar for the Eastern Wildway. Mike's talents served to make the Wildway look even more evocative and alive, and the watercolor paints by their very nature help emphasize that at the scale of the entire continent, the minute details and parcel-by-parcel pathways aren't as important as the big picture. Wildlands Network (and Reed Noss) had a history of combining art and deep green conservation theories in the pages of our *Wild Earth* magazine, and both the Florida Wildlife Corridor and Eastern Wildway art maps can be seen as growing out of that tradition.

Looking back on the last 14 years of habitat mapping work that I've been involved with here at Wildlands Network, it is pretty clear that none of it would have happened if Noss and Harris hadn't led the way and scouted the audaciously optimistic pathway forward. Our very existence as an organization can be traced to the bold ideas that Harris, Noss, and Hoctor put into print, that landscape-scale and even continental-scale conservation were both feasible and absolutely essential if we wanted to save the biodiversity of North America.

Literature Cited

Adirondack Council. 2020. 2020 Vision Series. https://www.adirondackcouncil.org/page/2020-vision-series-58.html.

Barborak, J. R., A. F. Carr III, and L. D. Harris. 1993. Recomendaciones para la consolidación territorial y conectividad de las areas protegidas de Costa Rica (Recommendations for the territorial consolidation and connectivity of protected areas in Costa Rica). Pages 113–133 in A. Vega, editor. Conservation Corridors in the Central American Region / Corredores Conservacionistas en la Región Centroamericana. Tropical Research and Development, Gainesville, Florida.

Bhutan Biodiversity. Undated. Protected Areas and Biological Corridors of Bhutan. Unpublished report.

Blackfoot Nation. 2020. Iinnii Buffalo Spirit Center. https://blackfeetnation.com/iinnii-buffalo-spirit-center/.
Carr, M. H., T. D. Hoctor, C. Goodison, P. D. Zwick, J. Green, P. Hernandez, C. McCain, J. Teisinger, and K. Whitney. 2002. Final Report: Southeastern Ecological Framework. Submitted to US Environmental Protection Agency, Region 4, Atlanta, Georgia. https://cfpub.epa.gov/roe/documents/sef_report_ROE.pdf
Cenozoic Society, The. 1992. The Wildlands Project: Plotting a North American Wilderness Recovery Strategy. Wild Earth: Special Issue. Canton, New York.
Chester, C. C. 2006. Conservation across Borders: Biodiversity in an Interdependent World. Island Press, Washington, DC.
Christy, A., and M. DiGirolamo. 2022. Wildlife crossings built with tribal knowledge drastically reduce collisions. Mongabay. 18 November. https://news.mongabay.com/2022/11/video-wildlife-crossings-built-with-tribal-knowledge-drastically-reduce-collisions/. Accessed September 5, 2024.
Dena Kayeh Institute. Undated. Dene K'éh Kusān. https://denakayeh.com/. Accessed October 10, 2023.
Foreman, D. 2004. Rewilding North America: A Vision for Conservation in the 21st Century. Island Press, Washington, DC.
Goldfarb, B. 2023. Crossings: How Road Ecology Is Shaping the Future of Our Planet. W. W. Norton & Company, New York.
Harvey, A. 1998. A sense of place: Issues, attitudes and resources in the Yellowstone to Yukon ecoregion. Yellowstone to Yukon Conservation Initiative, Canmore, Alberta, and Bozeman, Montana.
Hebblewhite, M., J. A. Hilty, S. Williams, H. Locke, C. Chester, D. Johns, G. Kehm, and W. L. Francis. 2021. Can a large-landscape vision contribute to achieving biodiversity targets? Conservation Science and Practice. https://doi.org/10.1111/csp2.588.
Hilty, J. A., A. L. Jacob, K. G. Trotter, M. Hilty, and H. Young. 2019. Endangered species, wildlife corridors, and climate change in the US West. In E. L. Wolters and B. Steel, editors. The Environmental Politics and Policy of Western Public Lands. Oregon State University, Oregon. https://open.oregonstate.education/environmentalpolitics/open/download?type=print_pdf#page=130
Hilty, J., and K. Zenkewich. 2022. The future of large landscape conservation begins with Indigenous communities. High Country News, November 1, 2022. https://www.hcn.org/issues/54-11/public-lands-the-future-of-large-landscape-conservation-begins-with-indigenous-communities/. Accessed September 5, 2024.
Laliberte, A., and W. Ripple. 2004. Range contractions of North American carnivores and ungulates. Bioscience 54:123–138.
Leopold, A. 1949. A Sand County Almanac. Oxford University Press, New York.
Locke, H. 1994. The Wildlands Project and the Yellowstone to Yukon Biodiversity Strategy: Preserving the wild heart of North America. Borealis 15:18–24.

Locke, H. 2013. Why nature needs half: A necessary and hopeful agenda for parks and protected areas. Parks: The International Journal for Parks and Protected Areas 19(2):9–18.

Locke, H. 2018. The International Movement to Protect Half the World: Origins, Scientific Foundations, and Policy Implications. Earth Systems and Environmental Sciences Reference Manual. Elsevier, Amsterdam, The Netherlands.

Locke, H., E. C. Ellis, O. Venter, R. Schuster, K. Ma, X. Shen, S. Woodley, N. Kingston, N. Bhola, B. B. N. Strassburgh, A. Paulsch, W. Williams, and J. Watson. 2019. Three global conditions for biodiversity conservation and sustainable use: An implementation framework. National Science Review 6:1080–1082.

Lundquist, L. 2019. UM Prof: Blackfeet tribe should co-manage Bager-Two Medicine with the Forest Service. Missoula Current, April 22, 2019. https://missoulacurrent.com/badger-two-medicine/. Accessed September 5, 2024.

MacArthur, R. H., and E. O. Wilson. 1967. The Theory of Island Biogeography. Princeton University Press, Princeton, New Jersey.

McKelvey, K. S., K. B. Aubry, B. Keith, N. J. Anderson, J. Neil, A. P. Clevenger, J. P. Copeland, K. S. Heinemeyer, S. Kimberley, R. M. Inman, J. R. Squires, J. S. Waller, K. L. Pilgrim, and M. K. Schwartz. 2014. Recovery of wolverines in the western United States: Recent extirpation and recolonization or range retraction and expansion? USDA Forest Service / UNL Faculty Publications 324. http://digitalcommons.unl.edu/usdafsfacpub/324. Accessed September 5, 2024.

Montana Fish, Wildlife, and Parks. 2023. Innovative program launched to reduce wildlife-vehicle conflicts. March 1, 2023. https://fwp.mt.gov/homepage/news/2023/mar/0301-innovative-program-launched-to-reduce-wildlife-vehicle-conflicts. Accessed July 7, 2023.

National Park Service. 2018. History of bison management in Yellowstone. Yellowstone National Park. US Department of the Interior. https://www.nps.gov/articles/bison-history-yellowstone.htm [last updated April 2014]. Accessed September 5, 2024.

Newmark, W. D. 1985. Legal and biotic boundaries of western North American national parks: A problem of congruence. Biological Conservation 33:197–208.

Nez Perce Tribe Water Resources Division (NPTWRD). 2019. Landscape level adaptation and conservation project planning. https://nptwaterresources.org/climate-change-program/blues-to-bitterroots-coalition/. Accessed September 5, 2024.

Nez Perce Wildlife Division. Undated. The Homecoming of Hîmiin. https://www.nezpercewildlife.org/gray-wolf.

Noss, R. F. 1985. Landscape considerations in reintroducing and maintaining the Florida panther: Design of appropriate preserve networks. 1985. Unpublished report to the Florida Panther Technical Advisory Council, Gainesville.

Noss, R. F. 1992. The Wildlands Project: Land conservation strategy. Wild Earth (Special Issue): 10–25.

Noss, R. F. 2003. A checklist for wildlands network designs. Conservation Biology 17:1270–1275.

Noss, R. F. 2012. Forgotten Grasslands of the South: Natural History and Conservation. Island Press, Washington, DC.

Noss, R. F., and A. Y. Cooperrider. 1994. Saving Nature's Legacy: Protecting and Restoring Biodiversity. Island Press, Washington, DC.

Noss, R. F., and L. D. Harris. 1986. Nodes, networks, and MUM's: Preserving diversity at all scales. Environmental Management 10:299–309.

Noss, R. F., W. J. Platt, B. A. Sorrie, A. S. Weakley, D. B. Means, J. Costanza, and R. K. Peet. 2015. How global biodiversity hotspots may go unrecognized: Lessons from the North American Coastal Plain. Diversity and Distributions 21:236–244. https://doi.org/10.1111/ddi.12278.

Odum, E. P., and H. T. Odum. 1972. Natural areas as necessary components of man's total environment. Transactions of the North American Wildlife and Natural Resources Conference 37:178–189.

Parks Canada. 2022. Report on the Plains Bison reintroduction pilot, 2017–2022. November 2022. Parks Canada. https://parks.canada.ca/pn-np/ab/banff/info/gestion-management/bison/rapport-reintroduction-report. Accessed September 5, 2024.

Proctor, M. F., W. F. Kasworm, K. M. Annis, A. G. MacHutchon, J. E. Teisberg, T. G. Radandt, and C. Seervheen. 2018. Conservation of threatened Canada-USA trans-border grizzly bears linked to comprehensive conflict reduction. Human-Wildlife Interactions 12:348–372.

Sagner, D. 2023. CSKT awarded $30 million to update and rehabilitate U.S. Highway 93. Flathead Beacon. https://flatheadbeacon.com/2023/06/16/cskt-awarded-30-million-to-update-and-rehabilitate-u-s-highway-93/. Accessed July 7, 2023.

Sartor, C. C., S. A. Cushman, H. Y. Wan, F. P. Tirelli, S. L. Althof, E. Eizirik, T. C. Trigo, and T. R. O. de Freitas. 2024. Prioritizing conservation areas to mitigate connectivity loss and local extinction risk of a small carnivore (*Leopardus guttulus*) in South America. Biodiversity and Conservation 33:1605–1622. https://doi.org/10.1007/s10531-024-02817-w.

Scott, T. 2023. Blackfeet bring bison home to Chief Mountain. Flathead Beacon, June 28, 2023. https://flatheadbeacon.com/2023/06/28/blackfeet-bring-bison-home-to-chief-mountain/. Accessed September 5, 2024.

Smith, D. W., D. R. Stahler, and D. R. MacNulty, editors. 2020. Yellowstone Wolves: Science and Discovery in the World's First National Park. University of Chicago Press, Chicago.

Soulé, M., and R. Noss. 1998. Rewilding and biodiversity: Complementary goals for continental conservation. Wild Earth 8(3):18–28.

Theobald, D. M., A. L. Jacob, P. R. Elsen, E. A. Beever, L. Ehlers, and J. A. Hilty. 2024.

Global mountain ecosystems are threatened by loss and fragmentation from human modification. Frontiers in Conservation Science. https://doi.org/10.1111/cobi.14240.
UNESCO. 1974. Task Force on criteria and guidelines for the choice and establishment of Biosphere Reserves. Man and the Biosphere Report No. 22. Paris, France.
Vega, A., editor. 1994. Conservation Corridors in the Central America Region / Corredores Conservacionistas Duncan la Region Centroamerica. Tropical Research and Development, Gainesville, Florida.
Weston, P., and P. Greenfield. 2022. COP15 2022: Trudeau pledges £510m of Indigenous-led conservation projects. Guardian News. December 8, 2022. https://www.theguardian.com/environment/2022/dec/08/cop15-trudeau-pledges-510m-for-indigenous-led-conservation-projects. Accessed July 7, 2024.
White, P. J., R. L. Wallen, C. Geremia, J. J. Treanor, and D. W. Blanton. 2011. Management of Yellowstone bison and brucellosis transmission risk—implications for conservation and restoration. Biological Conservation 144:1322–1334.
Wilson, E. O. 2016. Half-Earth: Our Planet's Fight for Life. Liveright, New York.
Yellowstone Science. 1994. The Yellowstone lion: Yellowstone Science 2:8–13. http://npshistory.com/publications/yell/newsletters/yellowstone-science/2-3.pdf. Accessed September 5, 2024.
Yellowstone to Yukon Conservation Initiative. 2020. Indigenous peoples leading the way on conservation in the Yellowstone to Yukon region and beyond. https://y2y.net/blog/indigenous-led-conservation-yellowstone-to-yukon-region. Accessed September 5, 2024.

Acknowledgments

At risk of leaving someone out, we humbly thank the following individuals and organizations for their advice, encouragement, practice, and funding of Florida's landscape conservation planning initiatives: Florida Natural Areas Inventory (Florida State University); The Nature Conservancy / Richard Hilsenbeck; Florida Department of Environmental Protection; Florida Fish and Wildlife Conservation Commission; US Fish and Wildlife Service; University of Kentucky / Dave Maehr; University of Florida; University of Central Florida; Florida Department of Transportation; Peninsular Florida Landscape Conservation Cooperative; Florida Greenways and Trails Council; Florida Greenways Commission; Florida Greenways Coordinating Council; Century Commission for a Sustainable Florida; Florida Conservation Group; Live Wildly; Bellini Better World; Florida Wildlife Corridor Foundation / Mallory Dimmitt, Jason Lauritsen; Carlton Ward / Wildpath; Archbold Biological Station / Joe Guthrie, Josh Daskin, Hilary Swain. We also thank two anonymous reviewers for their helpful comments; Liz Thompson (University of Florida Center for Landscape Conservation Planning) for preparing the final artwork in a consistent format; and our acquisitions editor, Janie Chan; managing editor, Marthe Walters; and Brianna Blackburn, project manager for Scribe Inc., for guidance, assistance, and helpful advice.

Contributors

AARON ADAMS is the director of science for the science-based nonprofit conservation organization Bonefish & Tarpon Trust. He is an author of more than 100 peer-reviewed research articles and the author of numerous books that translate ecological science into layman's terms, including *The Orvis Guide to Fly Fishing for Coastal Gamefish.*

EVE BOHNETT is an assistant scholar in the Department of Landscape Architecture at the University of Florida, where she is employed by the Center for Landscape Conservation Planning. Her PhD dissertation for the University of Florida addressed assessment of landscape biodiversity using complex spatial-temporal models.

ISABELLA GUTTUSO BROWNE, MLA, is the urban and recreational green infrastructure coordinator for the University of Florida's Center for Landscape Conservation Planning. She researches green infrastructure as a resiliency strategy in developed landscapes and the rural-urban interface, with a focus on integrating human and social factors into design and planning, particularly in the Florida Wildlife Corridor and broader landscape conservation efforts.

RAMESH BUCH has been focused almost entirely on the protection and acquisition of environmentally sensitive lands in Florida since 1991. He started with Miami-Dade County's Environmentally Endangered Lands program, then relocated to Gainesville in 2001 to run the Alachua County Forever Land Conservation Program. In 2016, he took the position of real estate services director for the St. Johns River Water Management District. He retired from public service in 2021 and joined the North Florida Land Trust as its director of conservation acquisitions. He has completed over 65,000 acres of property transactions worth over $232 million. He is a leader in assisting local governments with establishing their own land conservation programs.

MARGARET (PEGGY) CARR is a professor emerita of landscape architecture at the University of Florida. In 2006 she was recognized as a UF Research Foundation Professor. In 2007 she coauthored *Smart Land-Use Analysis: The LUCIS Model* with Paul Zwick. In 2008 she received Florida Blue Key's Distinguished Faculty Award. She served as a principal investigator overseeing in excess of $2 million in funded research for entities including the US Agency for International Development, US Environmental Protection Agency, Florida Department of Environmental Protection, St. Johns River Water Management District, and 1000 Friends of Florida.

SUSAN CARR is a program manager with the Putnam Land Conservancy, where she works with landowners and conservation programs to protect North Florida lands. Carr has a PhD from the University of Florida and an MS from Louisiana State University; her graduate research centered on the ecology of southeast US plant communities. Carr's work experience is varied, including roles with government and academic organizations and, more recently, nonprofit land trusts.

CHARLES CHESTER teaches global environmental politics at Brandeis University and is a coprincipal investigator of EMIGRA, an NSF-funded research group examining the role of migratory species in providing ecosystem services and equitable outcomes for people. He is currently building the website EarthWeb.info, a web guide to environment, conservation, and sustainability. He is a former chair of the Board of Bat Conservation International (BCI), currently sits on BCI's Science Advisory Committee, and is chair of the Yellowstone to Yukon (Y2Y) Council (and a former cochair of the Y2Y Board).

JOHN DAVIS is a director of the Rewilding Institute (rewilding.org) and rewilding advocate for the Adirondack Council (adirondackcouncil.org). He is a former editor of the *Earth First!* journal and *Wild Earth* magazine. At Adirondack Council, Davis strives to implement for Adirondack Park and surrounding areas the rewilding goals promoted by the Rewilding Institute.

LAUREN DAY is the Florida state director at The Conservation Fund. She has worked in land conservation for over 20 years, protecting more than 75,000 acres in Florida—from small neighborhood parks to vast working landscapes. Before joining the Conservation Fund in 2008, she served as the executive director of Alachua Conservation Trust, a nationally accredited nonprofit land trust. Day has a bachelor's degree in environmental studies from Eckerd College and a master's degree in real estate and urban analysis from the University of Florida.

MALLORY DIMMITT is the chief executive officer of the Florida Wildlife Corridor Foundation. Dimmitt brings knowledge and experience in landscape-scale and working lands conservation, environmental policy, water allocation, and non-profit management. Dimmitt has been a participant in all six Florida Wildlife Corridor Expeditions and films (2012–2023). She was formerly vice president of strategic development for Lykes Bros. Inc., a fifth-generation family-owned agribusiness based in Florida. Previously, Dimmitt led The Nature Conservancy's Colorado Plateau Initiative from Telluride, Colorado. Dimmitt was awarded a Doris Duke Conservation Fellowship at Duke University's Nicholas School of Environment, where she earned a master's of environmental management (MEM) in environmental economics and policy.

PAUL DONSKY is a hydrologist at the Suwannee River Water Management District and a PhD student at the University of Florida, where he is advised by Dr. Matt Cohen. He studies the ecology of Florida's springs. His master's thesis at the University of Florida was titled "Flood Disturbances Impact the Autotrophic Communities in the Karst Springs of the Suwannee River, FL."

RICHARD DOTY is the GIS coordinator and research demographer for the Bureau of Economic and Business Research at the University of Florida, which produces Florida's official state and local population estimates and projections. His primary area of expertise is in building GIS-based models for forecasting population and water demand using parcel-level data. His two most recent publications are *Florida Estimates of Population 2023* and *The Geospatial Small-Area Population Forecasting (GSAPF) Model Methodology Used by the Southwest Florida Water Management District.*

DANIEL FARRAH is the development and land-use analyst at the Center for Landscape Conservation Planning and an assistant scholar at the University of Florida. He has a bachelor's and master's degree in landscape architecture and a PhD in urban and regional planning. Farrah's current work focuses on future growth models in Florida, including the statewide Sea Level 2040/2070 study and individual county models.

ALAN FLINT is a retired research hydrologist from the US Geological Survey and currently codirector of Earth Systems Modeling for Earth Knowledge. He is a coauthor of several publications on refugia, including papers in *Climate Change Responses* and *Ecological Processes*.

LORRAINE FLINT is a retired research hydrologist from the US Geological Survey and currently codirector of Earth Systems Modeling for Earth Knowledge. She is a coauthor of several publications on refugia, including papers in *Frontiers in Ecology and the Environment* and *Trends in Ecology and Evolution*.

MARLY FULLER is the chief experience officer for the Florida Wildlife Corridor Foundation, bringing expertise in strategic engagement and communication. She holds an MBA from Virginia Commonwealth University and is dedicated to building meaningful connections with partners, donors, and communities to support the protection of the Florida Wildlife Corridor.

ASENA GOREN is a researcher at the Center for Landscape Conservation Planning at the University of Florida. She earned a PhD in Ecology at the University of Oslo and has over a decade of research experience spanning both academia and the public sector. Goren's work integrates scientific inquiry with practical applications, focusing on advancing landscape conservation strategies and fostering evidence-based decision-making to support sustainable environmental stewardship.

KEN GRUDENS has worked in the role of executive director with the Indian River Land Trust since 2004 to develop and implement a variety of programs that protect and conserve the natural resources and special places of Indian River County. Grudens has a master's degree in landscape architecture from the College of Environmental Science and Forestry in Syracuse, New York, and a bachelor's degree in architecture from the University of Buffalo. Grudens formally served as chair of the Indian River County Conservation Lands Committee and on an Advisory Committee of the Indian River Lagoon National Estuary Program. Currently, he serves on the Board of Directors of the Alliance of Florida Land Trusts.

JOSEPH M. GUTHRIE directs Archbold Biological Station's Predator-Prey Program, where he focuses on noninvasive techniques for wildlife research and road ecology. Guthrie leads the Corridor Observatory project, a collaborative long-term monitoring effort with the University of Florida to track wildlife occupancy and activity using camera traps and autonomous recording units in south-central Florida's working landscape. Prior to joining Archbold in 2020, Guthrie and colleagues from the University of Kentucky conducted research on the black bear population of Highlands and Glades counties from 2004 to 2012. This work inspired the Florida Wildlife Corridor campaign, which Guthrie co-led with Carlton Ward and Mallory Dimmitt. Guthrie worked as a conservation biologist for the Smithsonian Conservation Biology Institute in Front Royal, Virginia, from 2017 to 2020. He is a native of Henry County, Kentucky.

JODI HILTY is president and chief scientist of the Yellowstone to Yukon Conservation Initiative. Specializing in ecological corridor and large landscape research, she has more than 25 years of experience managing large landscape conservation efforts. This work focuses on applying best available information to address complex challenges through community-based, collaborative efforts and also centers Indigenous leadership in conservation. Author or editor of four books, she also led writing of the 2020 IUCN *Guidelines for Conserving Connectivity* as the deputy chair of the IUCN Connectivity Conservation Specialist Group.

THOMAS S. HOCTOR is director of the Center for Landscape Conservation Planning at the University of Florida. He is an expert on GIS applications for identifying conservation priorities and implementation actions for maintaining biodiversity and ecosystem services, including focal species habitat modeling, reserve design, wildlife corridors, expansion of protected lands to address climate change impacts, and strategies for ensuring effective conservation despite conflicts with land-use change. He has served as principal or coprincipal investigator on many regional-scale conservation analysis and planning projects, including the Florida Ecological Greenways Network and Florida Wildlife Corridor, the Florida Critical Lands and Waters Identification Project, Coastal Resilience for Florida Aquatic Preserves, and Southeast Regional Longleaf Pine Ecological Connectivity.

TOM KAY is the executive director of Alachua Conservation Trust. He is an attorney and serves on the boards of several local and regional conservation organizations.

AMY KNIGHT is the GIS program specialist for the Florida Natural Areas Inventory and has been the coprincipal analyst (along with Jon Oetting) for the Florida Forever Conservation Needs Assessment since 2001. She also has expertise in land-cover mapping, species habitat modeling, and conservation planning analyses for longleaf pine ecosystems. Knight has an MS in Zoology from the University of Florida and an MA in Anthropology from the University of Alabama.

JASON LAURITSEN is the chief conservation officer for the Florida Wildlife Corridor Foundation. With over 20 years of conservation expertise, he focuses on protecting vulnerable connections at risk of development in rapidly growing areas of the Corridor. He holds a master's in Science Education from the University of Iowa and a bachelor's in Animal Ecology from Iowa State University.

KAILYN LAWSON holds a master's degree in climate mitigation and adaptation from the University of South Florida and serves as the community conservation manager for the Florida Wildlife Corridor Foundation. In her role, she leads efforts to engage communities in conservation, creating impact through collaboration to preserve Florida's natural landscapes.

HARVEY LOCKE is a globally recognized pioneer in the field of connectivity and large landscape conservation. He is cofounder of the Yellowstone to Yukon Conservation Initiative and serves as its strategic adviser. He served as president of the Wildlands Project (the first international NGO dedicated to the field) and cofounder of the Nature Needs Half Movement and of the Nature Positive Initiative. He is also coauthor of IUCN's *Guidelines for Conserving Connectivity* and serves as vice chair for Nature Positive of the IUCN's World Commission on Protected Areas. He has received numerous conservation awards, has an honorary doctorate from the University of Calgary, and is a fellow of the Royal Canadian Geographic Society. He lives in Banff National Park, where his family has deep roots.

SARAH LOCKHART is a networking and communications analyst with the Center for Landscape Conservation Planning at the University of Florida. She has been with the center since 2018 and is working on her doctorate in interdisciplinary ecology at the School of Natural Resources and Environment, University of Florida. Lockhart received a master of environmental management degree from Duke University, where she focused on ecosystem science and conservation, and an undergraduate degree in marine science from Eckerd College. Her current research focus is on multiscalar connectivity conservation in Florida.

REED F. NOSS is an ecologist and conservation biologist, conservation science coordinator for the University of Florida's Center for Landscape Conservation Planning, and chief science adviser for the Southeastern Grasslands Institute. He has an MS in Ecology from the University of Tennessee and a PhD in Wildlife Ecology from the University of Florida. He has courtesy professor appointments at University of Florida and University of Central Florida. Noss is an elected fellow of the American Association for the Advancement of Science, former editor-in-chief of *Conservation Biology*, and past president of the Society for Conservation Biology. Among nearly 400 publications, his most recent books are *Forgotten Grasslands of the South: Natural History and Conservation* (2012) and *Fire Ecology of Florida and the Southeastern Coastal Plain* (2018).

MICHAEL O'BRIEN is the geographic information systems manager at the University of Florida's Center for Landscape Conservation Planning. He has worked on a variety of conservation planning projects, including the Florida Ecological Greenways Network, the Northwest Florida Wildlife Habitat Network, and the Southwest Florida Landscape Conservation Design. He has master's degrees in landscape architecture and geographic information systems.

JON OETTING is chief of conservation planning services at the Florida Natural Areas Inventory at Florida State University. Since the early 2000s Oetting has been a principal analyst on a variety of conservation planning efforts, including the Florida Forever Conservation Needs Assessment, the Critical Lands and Water Identification Project, and the Longleaf Sustainability Analysis. Oetting was also project editor for the *Atlas of Florida's Natural Heritage*, a comprehensive survey of Florida's natural resources and conservation efforts.

DEBI OSBORNE was a land conservation professional around the country for over 36 years, having worked most recently as director of land protection for the Conservation Foundation of the Gulf Coast for 11 years and previously as director of real estate for the National Audubon Society and field office director / project manager for the Trust for Public Land throughout the mid-Atlantic states. Osborne has assisted communities and public agencies with land conservation visioning as well as finance and stewardship issues and has conducted conservation training workshops at local, regional, and national conferences. Osborne holds a master's degree in city and regional planning from Harvard University and undergraduate degrees in social ecology and urban geography from the University of California, Irvine.

KATIE SHEPARD is cofounder of Essex Editions (essexeditions.com), a small artisanal publishing and author services company. In 2016, she published her first book, *Essex, New York Architecture: A Doodler's Field Guide*, cocreated with Geo Davis. Currently, she is the managing editor for *Rewilding Earth* (rewilding.org) and the box office and payroll manager for the Depot Theatre (depottheatre.org). She graduated from SUNY Plattsburgh with a bachelor's in English literature with minors in philosophy and Asian studies.

DANIEL J. SMITH is an assistant research scholar and member of the graduate research faculty in the Center for Landscape Conservation Planning, School of Landscape Architecture and Planning at the University of Florida. He is former chair of the National Academies Transportation Research Board Standing Committee on Environmental Analysis and Ecology and a member of the USFWS Florida Panther Recovery Program's Subteam on Transportation Issues. He has 35 years' experience in conservation ecology and environmental planning. His primary focus is studying movement patterns and habitat use of terrestrial vertebrates and integrating conservation, transportation, and land-use planning.

PHILIP STEVENS is a research scientist who oversees the Marine Fish Biology group at the Fish and Wildlife Research Institute of the Florida Fish and Wildlife Conservation Commission. The group focuses on reproductive biology, age and growth, habitat use, and movement patterns of marine fishes. He is an author of over 50 peer-reviewed publications on marine fish ecology, distribution and movements of aquatic animals, coastal restoration, and response of fish communities to natural disturbance and urban development.

RON SUTHERLAND serves as chief scientist and southeast program director for the conservation nonprofit Wildlands Network. Sutherland was the lead designer and architect for the Eastern Wildway, an ambitious map that presents a continental-scale vision for a wildlife habitat network from Florida to Quebec. Sutherland has also been involved in habitat connectivity, road ecology, and carnivore conservation projects across the US and Mexico, and he co-led a successful campaign to save the 79,000-acre Hofmann Forest in eastern North Carolina.

EMILY TAYLOR is an environmental scientist with the Water Resources and Natural Treatment Systems team at Jacobs Engineering and former director of the Howard T. Odum Florida Springs Institute. She holds a PhD in soil, water, and ecosystem sciences from the University of Florida, with expertise in aquatic ecosystem ecology and biogeochemistry. Her research focuses on how watershed development and land management practices influence water quality, nutrient dynamics, and ecosystem function in freshwater and coastal systems. Taylor has nearly a decade of experience working on water resource projects across Florida, including wetlands, springs, streams, and river systems. She has collaborated with government agencies, nonprofits, and community groups to support science-based environmental management and conservation.

JASON TEISINGER is a GIS project manager with the Bureau of Economic and Business Research at the University of Florida. He has dual bachelor of agriculture degrees in microbiology and botany and a master of arts degree in urban and regional planning with a GIS specialization from the University of Florida. He has 30 years of experience in building GIS-based population forecasting models using parcel-level data.

LIZ THOMPSON is a GIS analyst with a diverse background in urban and regional planning and K-12 education. She holds a bachelor of arts in geography and history from the University of Western Australia and a master's in urban and regional planning from the University of Florida. Certified to teach science and social studies for grades 6–12 in Florida, Thompson spent seven years teaching in the state's public schools. She also has over 15 years of GIS experience in government and academic institutions, specializing in regional urban planning, conservation planning, suitability analysis, and affordable housing research.

KIMBERLY TILLMAN is the partnership coordinator for Alachua Conservation Trust. She is a biologist who has focused her career on habitat restoration and management for threatened and endangered species. Throughout her career, Tillman has connected the public with conservation, led landscape-scale partnerships, and helped recover and improve the sustainability of plant and animal populations across the southeast US.

MICHAEL VOLK is the associate director of the Center for Landscape Conservation Planning at the University of Florida and a Research Assistant Professor in the Department of Landscape Architecture. Recent publications include Volk et al. 2018, Creating an initiative to meet climate change information needs for landscape architects (in C. Calorusso and T. Clements, eds., *Transforming the Discussion: Breaking our Landscape Architecture Chrysalis*); and Volk et al. 2018, Incorporating climate change into landscape architectural projects and practice (in N. Nawari and N. Clark, eds., *Tropical Storms as a Setting for Adaptive Development and Architecture*).

SONAM WANGDI is chief forestry officer at the Nature Conservation Division of the Department of Forests and Park Services of the Royal Government of Bhutan. As the chief of this division, he provides technical guidance and management advice to the protected areas of Bhutan. He has been the key person in declaring Bhutan's Ninth Biological Corridor that connects Bumdeling and Sakteng Wildlife Sanctuary in the eastern part of Bhutan, taking Bhutan's protected areas system to 52% of Bhutan's geographical area. He has also been instrumental in introducing the concept of Key Biodiversity Areas and in driving Other Effective Area–Based Conservation Measures (OECMs) in Bhutan.

CARLTON WARD JR. is a conservation photographer, Emmy Award–winning filmmaker, and National Geographic Explorer. He founded the Florida Wildlife Corridor project in 2010 and created Wildpath, an organization that uses storytelling to inspire the protection of wildlife corridors. Ward and Wildpath led the campaign to create the Florida Wildlife Corridor Act in 2021 and founded the Connected Planet project with National Geographic in 2025 to scale their model for accelerating the protection of wildlife corridors globally. Ward's work is featured in the National Geographic film *Path of the Panther*. He is an eighth-generation Floridian with a family ranch in the Peace River Valley and is now working to photograph and film the underappreciated beauty of the Gulf of Mexico.

SHANE WELLENDORF is the land conservancy director with Tall Timbers Research Station and Land Conservancy, an accredited land trust active throughout North Florida and South Georgia. The Land Conservancy works on conservation easements, landowner extension services, regional planning, and legislative advocacy. Wellendorf has been affiliated with Tall Timbers since 1996, working for many years as a research biologist with the Game Bird Lab and then transitioning to the Land Conservancy in 2011. Wellendorf holds a bachelor of science degree in wildlife biology from Iowa State University and a master of science degree in wildlife science from North Carolina State University and is a certified wildlife biologist with the Wildlife Society.

JOELLEN WILSON is the juvenile habitat program manager for the science-based nonprofit conservation organization Bonefish & Tarpon Trust. Her program focuses on juvenile sportfish habitat use, mapping and prioritization of nursery habitats, and integration of habitat into fisheries resource management.

DERYA YESILKUSAK is a PhD candidate in the Department of Landscape Architecture at the University of Florida. Her research focuses on sea-level rise, climate change, and climate adaptation. She is actively involved in projects such as the Coupled Model Intercomparison Project (CMIP) climate downscaling initiative and the Critical Lands and Waters Identification Project (CLIP) 5.0 stormwater management project, contributing to innovative solutions in climate resilience and landscape planning.

PAUL ZWICK is a professor emeritus of urban and regional planning at the University of Florida. In 2007 he coauthored *Smart Land-Use Analysis: The LUCIS Model* with Margaret Carr. In 2015 he authored *Advanced Land-Use Analysis for Regional Geodesign: Using LUCISplus* along with Iris E. Patten and Abdulnaser Arafat. During his tenure at UF, he served as the associate dean of the College of Design, Construction, and Planning; the department chair for the Department of Urban and Regional Planning; and the director of the GeoPlan Center.

Index

Note: Index includes frequently mentioned species (common names) and natural communities, as well as major place names and historical figures in this field.

1000 Friends of Florida, 38, 51, 60, 70, 73, 76, 79, 120

Acquisition and Restoration Council (ARC), 48, 60, 61
Adirondack, 212, 215; 2020, 212; Council, 212; Park, 212
Agriculture, 2, 24, 29–30, 32, 37–38, 62–63, 64, 76, *77*, 78, 81; Agriculture 2040/2070, 79, 81; Agriculture Conservation Easement Program–Wetlands Reserve Easement (ACEP-WRE) and Agriculture Land Easement (ACEP-ALE), 191. *See also* Easements: agricultural
Air Force Conservation Partnership, 133, 144
Alachua Conservation Trust (ACT), 192, 200, 204–205
Alachua County Forever (ACF), 192, 204
Algae, blue-green, 110
American Farmland Trust (AFT), 78–79, 82
America's Longleaf Restoration Initiative (ALRI), 186, 199–200
Amphibians, 15, 162, 165, 166, 175
Animal-vehicle collisions (AVC), 156, 158, 161–162, 164–169, 172–173, 175
Apalachicola National Forest, 36, 141
Aquatic: conservation, 3, 109–129; species, 15, 32, 91, 98, 113
Aquatic Preserves, 15, 55, 115
Aquifer, 53, 92, 96, 100, 113, 114, 119, 144; Biscayne, 101, 113; coastal karst, 100–101; Floridan, 113, 121
Archbold Biological Station, 140
Areas of Conservation Interest (ACIs), 49–51. *See also* Potential Natural Areas (PNAs)
Atlantic Warm Pool, 90, 91
Audubon Society, Florida (Audubon Florida), 49, 52

Babcock-Webb Wildlife Management Area, 47
Basin Management Action Plans (BMAPs), 114, 115, 116, 122
Bay, Florida, 109
Bay, Tampa, 113
Bear, Florida black, 26, 32, 38, 52, 59, 64, 100, 131, 133, 135, 140–141, 143, 159, 160, 164–169, 171

Bear, grizzly, 220, 221, 222
Bermuda High, 89
Best Management Practices (BMPs), 114, 116, 121
Bhutan, Kingdom of, 210, 217–218
Big Bend Coast (of Gulf of Mexico), 15, 37, 141, 203
Big Cypress National Preserve and region, 73, 146
Biodiversity: conservation, 2, 29, 34, 101, 102, 130–155, 157, 195, 197, 212, 218; hotspots, 1–2, 3, 5–21, 52, 89, 199, 211; loss, 5, 114, 223. *See also* Endemic species and Endemism
Biosphere reserve (UNESCO), 2, 213
Birds, 13, 32, 33, 100, 145, 156, 159, 165. *See also* Bobwhite, northern (quail); Caracara, crested; Crane, Florida sandhill; Owl, Florida burrowing; Scrub-jay, Florida; Woodpecker, red-cockaded
Bison (buffalo), 223
Blackwater River State Forest, 36; Blackwater-Conecuh conservation landscape, 150
Blue-green Algae Task Force (BGATF), 116
Bobwhite, northern (quail), 204
Bonefish, 110
Bottleneck planning, 150
Brock, Greg, 61
Buffer zones, 2, 14, 25, 35, 36, 37, 52, 55, 61, 74, 134, 144, 191, 205, 210, 213, 219, 223
Bureau of Economic and Business Research (BEBR), 71, 72, 73, 75, 76, 80
Burning, prescribed, 10, 200. *See also* Fire-dependent; Fire ecology

Cajeput, 95
Caloosahatchee Ecoscape, 64–65
Caloosahatchee River, 64, 109
Caracara, crested, 143, 145
Caves, 7, 96, 101
Carnivores (predators), 7, 13, 24–30, 140, 156, 162, 210, 218, 219, 221, 222; regulation of populations and food webs, 33; utility for conservation planning, 213. See also Panther, Florida
Cattle. *See* Livestock
Center for Landscape Conservation Planning, University of Florida, 14, 58, 60–63, 73, 75, 136, 139, 140, 151
Century Commission, 61
Choctawhatchee River Water Management Area, 64
Climate (and weather): stability/refugia in Coastal Plain, 8, 87, 88, 89–90, 101. *See also* Climate (climatic) refugium/refugia
Climate change, 75, 78–81, 82, 86, 87, 88, 92, 93, 95, 97, 98, 101, 102, 113, 123–124, 144, 222, 223. *See also* Endemic species and endemism
Climate (climatic) refugium/refugia, 3, 8, 86–108; microrefugia, 7, 87, 96–98, 101
Climate resilience/resiliency, 87, 88, 98, 123, 132, 146, 149
Climate velocity, 87
Closing the Gaps study/report, 50–51, 132
Coastal Barrier Resources Act, 133, 145
Coastal habitats/ecosystems, 15, 99, 100, 109, 111, 113, 117, 119, 123, 134
Coastal Plain, North American, 2, 7, 8, 10, 14, 87–90, 211. *See also* Biodiversity: hotspots; Climate (and weather)
Cogongrass, 157
Colonization (or recolonization), 9, 10, 30, 33, 34, 86; by humans, 5, 10. *See also* Dispersal
Common Ground Ecology, 62
Communities Trust, Florida, 47, 143, 187
Competition, 10

Connectivity, habitat or landscape, 13, 31–33, 49, 51–52, 58–60, 63–64, 76, 86, 101, 112, 115, 117–119, 122–123, 132, 134–136, 140–141, 144, 145, 151, 156, 157, 160, 164–167, 169, 171, 172, 175, 210, 218, 222, 224, 225
Conservation and Recreation Lands (CARL) program, 38, 47, 48
Conservation Area Design (CAD), 210
Conservation banks/banking. *See* Mitigation (conservation) banks/banking
Conservation biology, 1, 5, 13, 22, 28, 30, 130, 151, 157, 211. *See also* Conservation planning
Conservation Foundation for the Gulf Coast, 206
Conservation Fund, The (TCF), 38, 51, 187, 203
Conservation planning: as subdiscipline of conservation biology, 1, 5; systematic, 117, 118, 122, 186. *See also* Conservation biology
Convention on Biological Diversity (Global Biodiversity Framework), 28
Cooperative Conservation Blueprint, 61–62, 139
Cooperative Land Cover (CLC), FWC, 11, 173
Core/corridor/buffer concept, 2, 12, 14, 25, 29, 35–38, 49, 102, 210, 211, 213, 217–219, 222, 226
Corridors, 2, 3, 7, 10, 12, 14–15, 23, 26, 29, 31–33, 34, 35–38, 49, 51–54, 58–59, 61, 62, 64, 69–70, 80, 81, 100–102, 118, 130–155, 157, 167, 169, 172–175, 186, 196–198, 201–206, 210, 211, 213, 217–219, 220, 222–223, 225, 226. *See also* Connectivity, habitat or landscape; Florida Ecological Greenways Network, Florida Wildlife Corridor
Costa Rica, 213
Crab, blue, 115
Crab, stone, 96
Crane, Florida sandhill, 143, 145
Critical Lands and Waters Identification Project (CLIP), 61–62, 63, 133, 139–140, 144
Critical linkages, 58, 64, 136, 140, 141
Crocodile, American, 100

Dark Sky Parks, International, 146
Davis, M. C., 64
Department of Agriculture and Consumer Services, Florida (FDACS), 62, 63, 190
Department of Defense, US, 139, 190, 191
Department of Environmental Protection, Florida (FDEP) (formerly Florida Department of Natural Resources), 15, 16, 48, 52, 55, 58, 61, 64, 115, 116, 131, 139, 196, 201
Department of Transportation, Florida (FDOT), 49, 131, 140, 166–167, 173, 174, 175
Dimmitt, Mallory, 141, 142
Dispersal, 13, 30, 32, 86–87, 95, 134, 141, 163; barriers to, 32, 87; for colonizing patches, 32; in metapopulations, 32. *See also* Colonization (or recolonization); Metapopulations
Disturbance: as a contributor to high biodiversity, 10; changes with climate change, 96; creating shifting mosaics, 34; disturbance regime, 26, 27, 34; fire as, 10; human-caused changes in, 35; in relation to resistance and resilience, 96, 98; stimulating resprouting and growth of plants, 10; to soil, 10
Diversity, beta, 13–14; nestedness vs. turnover patterns, 13–14. *See also* Biodiversity

Driver-Pressure-State-Impact-Response (DPSIR), 119–120
Drought: increasing with climate change, 95, 124
Dry prairie, Florida (hyperseasonal subtropical grassland), 144, 145; restoration of, 148, 227
Drum, red, 110, 112, 122
Ducks Unlimited, 184

Easements, 145, 148, 201, 212; advantages and disadvantages of, 149, 193–196; agricultural, 62, 64, 139, 145, 147, 190–191, 198, 201; conservation, 12, 16, 38, 48, 62, 64, 75–76, 141, 145, 148, 149, 182, 183–184, 186, 187–191, 193–196, 198, 199, 200, 201, 204, 206, 212; Forever Wild, 212; history, 183–184; Wetland Reserve Easement program, 191, 206. *See also* Land trusts
Earth First! (journal and group), 27, 28
Eastern Wildway, 2, 150, 224, 226–227
Ecological engineers, 33. *See also* Herbivores, large
Ecological Society of America, 24
Ecologists' Union, 25–26
Ecosystem, 6, 7, 10, 13, 15, 23–26, 27, 33, 53, 93, 95, 99, 101, 113, 114, 116–117, 120, 122, 123, 124, 134, 139, 146, 151, 157, 162, 164–165, 166, 169, 171, 182, 195, 199, 204, 213, 221, 222, 223; definition of by Tansley, 23
Ecosystem services, 2, 7, 14, 28–29, 64, 101, 121, 130–155, 201, 218
Eglin Air Force Base, 36, 64, 141
Endemic species and endemism, 2, 6–7, 12–13, 87–88, 98, 102, 210; allopatric speciation of on ridges, 8, 89; association with regions with high climatic stability, 8, 88–89; concentration in southeastern Coastal Plain and fire-prone ecosystems, 7–8, 10. *See also* Climate (and weather): stability/refugia in Coastal Plain
Environmentally Endangered Lands (EEL) program, 38, 47, 48
Estuaries, 15, 55, 95, 96, 98, 110–111, 113, 114, 115, 117–124, 146, 148
Everglades National Park (and region), 15, 57, 99, 100, 101, 109, 141, 142, 146, 148, 190, 191
Everglades Stormwater Treatment Areas (STAs), 15, 99, 100, 101, 109, 142, 148, 190, 191
Everglades to Gulf Conservation Area, 190
Evolutionary cradle and museum, 8
Evolutionary environment (and evolutionary fire ecology/regime/season), 10
Exotic species. *See* Nonnative species

Farm Bill, US, 190, 191, 198
Fire ants, 157
Fire-dependent (species, communities, ecosystems), 10, 199
Fire ecology, 10–11, 143, 195, 199
Fish and Wildlife Conservation Commission, Florida (FWC) (formerly Florida Game and Fresh Water Fish Commission), 16, 50, 51, 61, 62, 110, 112, 116, 132, 133, 139, 140, 143, 144, 161, 166, 168, 169, 173, 193
Fish and Wildlife Service, US (USFWS), 120, 133, 139, 164, 165, 169, 173, 190, 203
Flatwoods, 32, 134, 144. *See also* Savanna, pine
Flood(ing): as part of disturbance regime and agent of selection, 7, 96, 118, 119, 132, 143–145, 149, 163; increasing with climate change, 98. *See also* Hydroperiod

Florida 2060, 73, 75, 76
Florida 2070, 3, 76, 77, 80
Florida Bay, 109
Florida Communities Trust (FCT), 47, 143, 187
Florida Conservation Group, 150, 184
Florida Ecological Greenways Network (FEGN), 3, 6, 8, 12, 38, 52, 58, 61, 64, 69, 130, 144, 145, 147, 149, 151, 167, 186, 196, 225, 226. *See also* Florida Wildlife Corridor
Florida Forever, 5, 7, 46–48, 52–57, 59–62, 64, 76, 77, 80, 133, 135–137, 139–141, 143, 147, 150, 187, 189, 201, 212; Act/Program, 48, 52, 54–57, 61, 141, 187; Conservation Data Viewer, 61; Conservation Needs Assessment (FFCNA), 55–57, 60–63; Expert Working Group, 60
Florida Keys, 8, 99; and endemic species, 8
Florida model (of conservation planning), 2, 3, 6, 9, 23, 210–214, 216–219
Florida National Scenic Trail, 146
Florida Natural Areas Inventory (FNAI), 14, 48–50, 55–58, 60–62, 64, 75, 89, 96, 132, 133, 135, 136, 140, 143, 144, 150, 190, 193
Florida Springs Protection Task Force (FSPTF), 115, 116
Florida Wildlife Corridor, 3, 5–8, 12, 14, 15, 23, 26, 29, 36–38, 52, 54, 58, 60, 61, 62, 64, 69, 80, 81, 101, 118, 130, 131, 133, 135, 137, 139–149, 151, 152, 167, 169, 172, 173, 175, 186, 196, 197, 201, 202, 203, 205, 225–227; Florida Wildlife Corridor Act; Florida Wildlife Corridor Foundation, 58, 142, 146–148, 150, 186, 201. *See also* Florida Ecological Greenways Network
Floridan Aquifer Collaborative Engagement for Sustainability (FACETS), 121
Forbes, Stephen, 23
Foreman, Dave, 28, 210, 211, 214, 215, 225
Forest Legacy Program, 148, 190
Forest Service, US, 148, 190, 203, 222
Forestry, 2, 57, 149, 152, 194. *See also* Agriculture
Forgotten Grasslands of the South (book), 12, 211, 241
Forman, Richard T. T., 34, 64, 156, 157, 161, 175
Fossils, 10, 90
Foundation species, 121, 122
Fragmentation, habitat or landscape, 12, 30, 31, 35, 51, 60, 109, 111, 115, 130, 131, 138, 143, 156, 157, 165, 169, 175
F-TRAC, 55, 57–60
Functional Wetlands, 145

Geographic information systems (GIS), 22, 28, 34, 50, 52, 55, 61–63, 70–72, 82, 131, 132, 151, 159, 160, 161, 167, 172, 187, 211, 225–227
Geology, 101
GeoPlan Center (University of Florida), 73, 76, 151, 167, 174, 245
Glacial-interglacial (and stadial-interstadial) cycles, 87, 88, 90, 91. *See also* Last Glacial Maximum
Gopher tortoise. *See* Tortoise, gopher
Grazing, 145, 191, 194, 210. *See also* Herbivores, large; Livestock
Green infrastructure, 14, 63, 76, 149
Green Swamp Management Area (Florida), 73, 142
Greenways Commission, Florida, 38, 51, 69, 131, 203
Greenways Plan/Program, Florida, 38, 51, 52, 131, 138, 139
Groundwater, 53, 97, 98, 113, 114, 210
Gulf Coastal Corridor, 10

Gulf of Mexico, 37, 89, 90, 100, 113, 198
Gulf Stream (and Florida Current, Loop Current), 90, 91
Gulf to Atlantic Restoration Partnership (GARP), 200
Guthrie, Joe, 130, 140–142

Half-Earth (book), 29
Hammock (hardwood forest): 16, 144
Hardwood forest. *See* Hammock
Harris, Larry, 2, 12, 32–37, 51, 131, 164, 203, 210, 212, 224–227
Healthy Farms Healthy Bays (HFHB), 121
Herbivores, large, 28, 33, 223
High pine. *See* Sandhill (and clayhill, high pine); Savanna, pine
Hilsenbeck, Richard, 233
History: of the Florida Ecological Greenways Network and Florida Wildlife Corridor, 130–143; of land acquisition programs in Florida, 47–63; of land trusts and landscape conservation in Florida, 186–187; of regional landscape conservation planning, 22–38; of road ecology research and practice in Florida, 164–171
Hoctor, Thomas, 52, 131, 140, 211, 212, 225, 227
Holocene Epoch, 92. *See also* Fossils; Humans
Homosassa River, 115
Hotspots (of biodiversity, endemism), 3, 5–7, 9, 11, 13, 15, 52, 89, 199, 211; significance to conservation, 160, 172, 173. *See also* Biodiversity
Hubs, 52, 58, 132, 134, 135, 136, 141, 173
Humans: land use and other activities, 2, 6, 10, 11, 13, 68, 70–75, 77, 78, 80, 82, 83, 111, 114, 124, 131, 151, 162, 165, 184, 187, 209, 223. *See also* Population growth (human)
Hurricanes, 10, 95, 96, 98, 221. *See also* Wind
Hydrology, 7, 96, 98, 101, 109, 110, 111, 118, 119, 132, 143–145, 149, 157, 163. *See also* Flood(ing)
Hydroperiod, 163, 171

Immigration (to Florida), 5, 32
Incentive strategies and programs, 65, 148, 194, 200
Indian River Lagoon Blueway, 205
Indian River Land Trust (IRLT), 192, 205
Indigenous People, 219, 220, 223
Inflation Reduction Act of 2022, 201
Insects, 172
Integrated Coastal Zone Management (ICZM), 117
Integrated Habitat Network (IHN), 135
Integrated Land-Sea Management (ILSM), 117
Integrated River Basin Management (IRBM), 117
Internal Revenue Service (IRS), 82, 183, 195
Intertropical Convergence Zone, 90
Invasive species, 95. *See also* Nonnative species
Irreplaceability vs. vulnerability, 6, 7, 11, 72, 98, 100, 119
Island biogeographic theory, 13, 30, 31
Island Press, 212, 216

Jefferson, Thomas, 219
Jenkins, Robert E., 26, 48
Jonathan Dickinson State Park, 36

Karst, 92, 96, 100, 101

Keystone (species or community/ecosystem), 30, 33, 223
Kissimmee Prairie (Preserve State Park), 146
Kissimmee River Basin, 191

Lake Okeechobee, 109, 110
Lake Tulane, 90, 91
Lake Wales Ridge, 90, 101, 142
Land and Water Conservation Fund, 190, 203
Landscape Conservation Design (LCD), 117, 120
Landscape ecology, 22, 34, 35, 130, 151, 226
Landscape linkages, 36, 53, 131, 172
Landscape species, 133, 143
Land Trust Alliance, 185, 188, 195
Land trusts, 3, 11, 78, 148, 149, 182–193, 195–206, 225, 226; and conservation partnerships, 198–199; as facilitators, 197–198; Land Trust Accreditation Commission, 188
Large, connected landscapes vs. small, isolated sites, 2, 12, 15, 102
Last Glacial Maximum, 90. *See also* Glacial-interglacial (and stadial-interstadial) cycles
Least-cost path (and corridor) models, 172–174
Livestock: burning to provide forage for, 77; effects on vegetation and fire regimes, 30. *See also* Herbivores, large; Humans
Live Wildly Foundation, 61. *See also* Florida Wildlife Corridor
Local Implementation Teams, 200
Loop Current, Gulf, 90, 91. *See also* Gulf Stream (and Florida Current, Loop Current)
Lower Suwannee National Wildlife Refuge, 203
Loxahatchee National Wildlife Refuge, 36

Mammals. *See* Herbivores, large; Holocene Epoch; Livestock; Megafauna extinction; Miocene Epoch; Panther, Florida; Pleistocene Epoch; Pliocene Epoch; Squirrel, fox
Manatee (Florida), 98, 144
Mangrove/mangroves (swamp), red, 112, 122
Map Viewer and Dashboard, 61, 151; FEGN Map Viewer, 3, 5, 12, 14, 15, 23, 26, 29, 38, 52, 58, 59, 61, 69, 70, 71, 72, 77, 80, 130–140, 143–147, 149–152, 167, 174, 186, 196, 201
Marl prairie, 144
Marsh, 122, 144. *See also* Swamp; Wetlands
Marxan (site selection algorithm), 3, 57, 58, 64
Matrix (landscape, natural communities), 7, 30, 35, 144
Megaherbivores. *See* Herbivores, large
Megafauna extinction, 222. *See also* Herbivores, large
Megalinkages, 211
Metapopulations, 31, 33
Minimum Flows and Levels (MFLs), 114, 115
Miocene Epoch, 8, 10, 89
Miami Rock Ridge, 101
Microrefugia, climate. *See* Climate (climatic) refugium/refugia
Mitigation (conservation) banks/banking, 192
Multiple-use module (MUM), 2, 35
Myakka River, 206

Nassau County, 186
National Bobwhite Conservation Initiative, 186
National Climate Assessment, 5th, 92
National Estuarine Program (NEP), 118
National Oceanic and Atmospheric Administration (NOAA), 100
National Park Service, 24, 183, 223
National Resource Conservation Service (NRCS), 77, 139, 148, 190, 191, 198, 199, 206; Regional Conservation Partnership Fund (RCPP)
National Wild and Scenic Rivers, 55
National Wildlife Refuges (NWRs), 36
Natural community, definition. *See names of individual natural community types (e.g., Dry prairie, Flatwoods)*
Natural heritage program (network), 26, 48
Natural Floodplains, 133, 144, 145
Natural range of variability, 93, 94, 172
Nature Conservancy, The (TNC), 26, 48, 49, 52, 70, 88, 183, 184, 203, 216
Nature Needs Half, 29, 210, 211, 213, 218, 226
NatureServe, 9
Neogene Period. *See* Miocene Epoch; Pliocene Epoch
Nestedness. *See* Diversity, beta
New Mexico Highlands, 211
Niche, 22, 32, 95, 101
Nonnative species, 14, 95, 199
North American Wetlands Conservation Act, 190
North Florida Land Trust (NFLT), 186, 192, 198, 205
Noss, Reed, 28, 34–37, 42, 49, 51, 70, 131, 210–213, 217, 218, 224–226
Nutrient(s), 110, 112, 114, 116, 122, 123, 150. *See also* Productivity (primary)
Ocala to Osceola Wildlife Corridor (O2O), 142, 198, 205
Ocala National Forest, 32, 36, 72, 73, 142, 159, 161
Odum, Eugene P., 29, 218
Odum, Howard T., 29, 98, 218
Office of Greenways and Trails, 52, 139
Ohio Department of Natural Resources, 36
Okefenokee National Wildlife Refuge, 36, 131, 142
Online planning tools. *See* Map Viewer and Dashboard
Open space preservation, 195
Opportunity areas, 29, 149, 201. *See also* Florida Wildlife Corridor
Osceola National Forest, 36, 142, 203
Outdoor Recreation and Conservation Program, 47
Outstanding Florida Waters, 55, 134
Owl, Florida burrowing, 9, 145

Panther, Florida, 7, 11, 26, 33, 38, 52, 59, 64, 100, 131, 133, 135, 140, 142, 143, 164–169, 172–174, 213, 224
Parsons, Dave, 211
Partington, Bill, 37
Partnership for Gulf Coast Land Conservation (PGCLC), 198
Paseo Pantera, 213, 217
Pastures (improved and semi-improved, importance of), 64, 145
Patch dynamics, 34
Path of the Panther (project, film), 142, 244
Paynes Prairie (Preserve State Park), 165, 166, 171
Pelican Island National Wildlife Refuge, 47
Peninsula effect, 90

Peninsular Florida Landscape Conservation Cooperative (PFLCC), 75
Pepper, Brazilian, 95
Phylogenetic reconstruction/studies/evidence, 8
Pine(s), 32, 64, 90, 134, 144, 199, 224, 227; fossils of and history, 90. *See also* Flatwoods; Sandhill (and clayhill, high pine); Savanna, pine
Pine, longleaf, 14, 64, 96, 150, 151, 199, 204, 227
Pine rockland, 14
Pine savanna. *See* Flatwoods; Sandhill (and clayhill, high pine); Savanna, pine
Pinhook Swamp, 36, 203
Plants. *See specific name or group name (e.g., grasses, oaks, palms)*
Pleistocene Epoch, 88, 89
Pliocene Epoch, 88, 89. *See also* Pleistocene Epoch
Pollen, fossil, 90
Population growth (human), 6, 11, 68–75, 77, 78, 80, 82, 83, 114, 124, 131, 151, 184, 187
Population models (spatial), 22, 34, 68, 73
Potential Natural Areas (PNAs), 50, 132–134. *See also* Areas of Conservation Interest (ACIs)
Prairie. *See* Dry prairie, Florida; Marl prairie; Wet prairie
Precipitation (rainfall): climate change and, 89–92, 94–96, 98, 99, 124, 209. *See also* Climate (and weather)
Prescribed burning. *See* Burning, prescribed
Preservation 2000 (P-2000), 38, 46, 47, 49, 51, 52, 64, 197
Prioritization, 3, 11, 47, 49, 51, 52, 55, 57, 59, 61, 63, 70, 71, 75, 131, 135, 136, 138, 197, 215
Priority Ecological Areas (PEAs), 132, 133, 144–146
Productivity (primary), 96, 117

Quail. *See* Bobwhite, northern (quail)
Quaternary Period, 96. *See also* Pleistocene Epoch; Pliocene Epoch

Rainfall. *See* Precipitation
Ranching/ranchland, 79, 145, 149, 152. *See also* Agriculture
Readiness and Environmental Protection Integration (REPI) program, 191, 205. *See also* Department of Defense; Sentinel Landscapes Partnership program
Recreation, resource-based, 145, 146
Red Hills, 150, 188, 204
Refugia. *See* Climate (climatic) refugium/refugia
Regime (state) shift, 10, 115, 163
Regional Conservation Partnership Program (RCPP), 148, 191, 198. *See also* Easements; National Resource Conservation Service
Reptiles (herpetofauna, in part), 162, 165, 166, 171, 175
Rescue effects (demographic and genetic), 32
Resilience (and resistance), 14, 15, 87, 88, 98, 101, 110, 123, 132, 134, 146, 149. *See also* Climate change
Resist-Adapt-Direct (RAD), 123
Restoration: 12, 22, 26–28, 33, 48, 53, 55, 62, 100, 115, 117, 118, 120–123, 135, 148, 186, 190, 199, 200, 209, 211, 222, 223
Rewilding, 26–30, 33, 37, 210, 211, 213–215, 221, 223; *Rewilding North America* (book), 211. *See also* Wilderness: recovery areas
Roads: road crossing locations, 159, 173,

174; road ecology, 3, 156, 157, 159, 161, 163–167, 169, 171, 173, 175; roadkill, 158, 160, 161, 164, 166, 168; roadkill hotspots, 158; roadside animal detection systems (RAD), 123, 167. *See also* Animal-vehicle collisions; Wildlife crossings (wildlife crossing structures)
Ross Prairie, 171
Rural and Family Lands Protection Program (RFLPP), 46, 62, 63, 120, 138–141, 147, 150, 190, 201

Salt marsh, 99. *See also* Marsh
Sandhill (and clayhill, high pine), 64, 143–145
Sand skink, Florida, 193
Santa Fe River, 204
Savanna, pine, 10, 96, 227
Save Our Coasts, 47
Save Our Rivers, 38, 47
Saving Nature's Legacy (book), 211
Science-implementation gap, 5, 6
Scrub, 10, 16, 32, 134
Scrub-jay, Florida, 9, 134, 193
Seagrass, 55, 109–112, 122
Sea level: Florida Keys and, 58, 60, 98, 99, 100; high rate of rise in early Holocene, 92; ridges in Florida insular during high stands, 8, 88, 89; rising with current global climate change, 78, 80, 87, 88. *See also* Climate change
Sea Level 2040 and Sea Level 2070, 79–81
Sea Level Rise Biodiversity Assessment, 75
Season (dry and wet): flooding and effects on plants, 10, 28, 143, 163; pattern of dry and wet in Florida, 89
Seminatural (landscape or fire regime), 7, 11, 13, 49, 60, 64, 82, 134, 135, 145
Seminole Tribe Indian Gaming Revenue Trust, 148
Sentinel Landscapes (partnership program), 139, 191. *See also* Department of Defense; Readiness and Environmental Protection Integration (REPI) program
Seven Runs Creek Conservation Easement, 64
Shelford, Victor, 23–27; goal of committees of the Ecological Society of America, 24
Silver Springs, 98
Silviculture, 145. *See also* Agriculture
Sky Islands, 211, 215
Sinkholes, 7, 96, 98
Snook, common, 110–113, 115, 118, 122
Society for Conservation Biology, 28
Socioeconomic(s), 114, 121
Soulé, Michael, 5, 14, 28, 29, 33, 210, 213, 226
Southern Rockies, 215
Spatial population models (modeling), 68, 73
Species richness, 10, 32, 35, 95. *See also* Biodiversity
Springs (freshwater), 36, 53, 55, 92, 97, 98, 101, 112–117, 142, 146, 166; Outstanding Florida, 55, 116, 134
Squirrel, fox, 143
State Wildlife Action Plan (SWAP), 186
St. Johns River Water Management District, 71
St. Marks National Wildlife Refuge, 36
Stoltzfus, Elam, 141
Strategic Habitat Conservation Areas (SHCAs), 51, 53, 55, 133, 144
Stressors. *See* Threats
Suncoast Parkway, 169, 171
Suwannee River, 36, 115, 121

Swamp, 144
Systematic conservation planning, 118, 122, 186

Tall Timbers (Research Station and Land Conservancy), 200, 204
Tampa Bay, 113
Tansley, Arthur, 23
Tarpon, 115
Technical Advisory Group (TAG), 52, 62, 131, 134, 138
Terrafirma, 188
Threats, 7, 51, 60, 63, 78, 81, 82, 100, 156, 206; in Florida, 6, 7, 98, 100, 110
Three Global Conditions for Biodiversity Conservation and Sustainable Use, 218
Tide, red, 110, 112
Tidal creek, 111, 118
Tiger, 171, 217, 218
Tigrina, 218
Tompkins, Doug, 28
Torreya State Park, 47
Tortoise, gopher, 33, 134, 171, 193; management plan, 171, 193; relocations and recipient sites, 193
Total Maximum Daily Loads (TMDL), 114
Traffic calming measures, 162
Troll, Carl, 34
Trophic cascades, 33, 222. *See also* Carnivores (predators); Keystone
Trust for Public Land, 183, 192
Turnover, 31. *See* Diversity, beta
Turtle, loggerhead sea, 96

Underrepresented natural communities, 133, 144
UNESCO. *See* Biosphere Reserve (UNESCO)
University of Florida, 14, 29, 34, 51, 52, 58, 71–73, 75, 79, 131, 133, 136, 139, 140, 150, 204; Conservation Clinic, 204
University of Maryland, 133, 135
Upland hardwood forest, 144
Urban, 5, 11, 13, 14, 30, 53, 72, 74, 77, 78, 96, 111, 113, 119, 120, 194; green infrastructure modeling, 14, 63, 76, 149
Urbanization, 75, 76, 111, 114, 121

Ward, Carlton, Jr., 130, 140–142
Water/waters, 15, 28, 38, 50, 52–55, 60, 62, 64, 68, 71, 72, 75, 77, 90, 91, 95–98, 101, 109–119, 121, 124, 133, 134, 138, 140, 143–146, 148, 149, 190, 192, 194, 203, 205, 210. *See also* Hydrology; Hydroperiod; Precipitation (rainfall)
Water Management Districts (WMDs), 38, 47, 71, 115, 206; Northwest Florida, 115; South Florida, 115; Southwest Florida, 115; Suwannee River, 115; St. Johns River, 71, 115
Watershed, 27, 101, 109, 111, 112, 115, 117–121, 123, 150, 206, 220; conservation, 101, 111, 117–122, 123, 144; management, 27, 112–117, 150, 220
Watershed Landscape Conservation Framework, 117
Weather, 89, 95, 96, 101. *See also* Climate (and weather)
Wekiva: Parkway, 169, 171; River, 36, 73
Wekiwa Springs State Park, 36, 166
Wetlands, 10, 15, 57, 64, 71, 122, 124, 133, 134, 144–146, 164, 190, 191, 206
Wet prairie, 166
Wild Earth (magazine), 28, 212, 213, 227
Wilderness, 7, 16, 25–30, 210–212, 214–216, 221, 224; enthusiasm of public, 7: wilderness recovery areas, 27. *See also* Rewilding
Wildlands Conservation, 184

Wildlands Project (Network), The, 28, 210–216, 224–227
Wildlife crossings (wildlife crossing structures), 11, 35, 163–165, 167, 169, 171. *See also* Roads
Wildways, 2, 150, 211, 215, 224, 226, 227; Continental, 211; Eastern, 2, 150, 226–227; Great Plains, 211; Gulf Coast, 211
Wilson, E. O., 29
Wind, 10, 221
Withlacoochee State Forest, 47
Wolf, 222, 224; Gray, 217; Mexican, 211
Woodpecker, red-cockaded, 96
Woodville Karst Plain, 100
Wright, George Melendez, 25, 27

Xeric habitat, 134, 144

Yellow River Ravines, 36
Yellowstone to Yukon Conservation Initiative, 210, 217–219
Younger Dryas, 90

FSC
www.fsc.org
MIX
Papier | Fördert gute Waldnutzung
FSC® C083411

Kolibri 360 GmbH
Ferdinand-Jühlke-Straße 7
99095 Erfurt, Deutschland
produktsicherheit@kolibri360.de